D1000343

Probabilistic Programming

Probability and Mathematical Statistics

A Series of Monographs and Textbooks

Editors

Z. W. Birnbaum
University of Washington
Seattle, Washington

E. Lukacs
Catholic University
Washington, D.C.

---

1. Thomas Ferguson. Mathematical Statistics: A Decision Theoretic Approach. 1967
2. Howard Tucker. A Graduate Course in Probability. 1967
3. K. R. Parthasarathy. Probability Measures on Metric Spaces. 1967
4. P. Révész. The Laws of Large Numbers. 1968
5. H. P. McKean, Jr. Stochastic Integrals. 1969
6. B. V. Gnedenko, Yu. K. Belyayev, and A. D. Solovyev. Mathematical Methods of Reliability Theory. 1969
7. Demetrios A. Kappos. Probability Algebras and Stochastic Spaces. 1969, 1970
8. Ivan N. Pesin. Classical and Modern Integration Theories. 1970
9. S. Vajda. Probabilistic Programming. 1972
10. Sheldon Ross. Introduction to Probability Models. 1972
11. Robert B. Ash. Real Analysis and Probability. 1972
12. V. V. Federov. Theory of Optimal Experiments. 1972

In Preparation

Tatsuo Kawata. Fourier Analysis in Probability Theory

Fritz Oberhettinger. Tables of Fourier Tranforms and Their Inverses

H. Dym and H. P. McKean. Fourier Series and Integrals

PROBABILISTIC PROGRAMMING

S. VAJDA

THE UNIVERSITY OF BIRMINGHAM
BIRMINGHAM, ENGLAND

1972

ACADEMIC PRESS New York and London

Copyright © 1972, by Academic Press, Inc.
ALL RIGHTS RESERVED
NO PART OF THIS BOOK MAY BE REPRODUCED IN ANY FORM, BY PHOTOSTAT, MICROFILM, RETRIEVAL SYSTEM, OR ANY OTHER MEANS, WITHOUT WRITTEN PERMISSION FROM THE PUBLISHERS.

ACADEMIC PRESS, INC.
111 Fifth Avenue, New York, New York 10003

United Kingdom Edition published by
ACADEMIC PRESS, INC. (LONDON) LTD.
24/28 Oval Road, London NW1 7DD

Library of Congress Catalog Card Number: 72-180796

AMS(MOS) 1970 Subject Classifications: 62-02, 62C25, 90-02, 90A15, 90B30, 90C15

PRINTED IN THE UNITED STATES OF AMERICA

Contents

II. Decision Problems

III. Chance Constraints

Appendix I

Appendix II

Introduction

Mathematical programming (i.e. planning) is an application of techniques to the planning of industrial, administrative, or economic activities. Analytically, it consists of the optimization (maximization or minimization) of a function of variables (the "objective function") which describe the levels of activities (production of a commodity, distribution of facilities, etc.) and which are subject to constraints (e.g. restrictions in the availability of raw material, or on the capacities of communication channels). As a rule, the variables are also restricted to taking only nonnegative values.

The formulation of applied problems will incorporate "technological" coefficients (prices of goods, cost and capacity of production, etc.) on which a model of the situation to be analyzed can be based. In the classical situation these coefficients are assumed to be completely known. But if one wants to be more realistic, then this assumption must be relaxed. Tintner (1941) distinguishes between subjective risk, when "there exists a probability distribution of anticipation which is itself known with certainty," and subjective uncertainty, when "there is an a priori probability of the probability distributions themselves."

The former field leads to stochastic, or probabilistic, programming.

We assume, then, that the (joint) distribution of the technological coefficients is given. This includes the case when they are independently distributed, and cases when some of them have given values, without stochastic deviations.

A number of possible attitudes to deal with such a situation have been proposed. For instance, we might be ready to wait until the actual values of the coefficients and constants become known—for instance the requirement for some commodities during the next selling period—but we find that we must choose now in which type of activity we shall invest funds which we have available.

Under these circumstances we shall remind ourselves that if the coefficients are random variables, then the best result which we can obtain, after their values have become known and are taken into account to find the most favorable activities, is also a random variable.

Assume, then, that various investment possibilities have been offered to us, all of them incurring some risk, but that we are confident that we shall be able to find the best procedure in all possible emerging situations. Which criteria shall we then apply in choosing among the offers?

This is a problem for economists not for mathematicians. Economists tell us that we might choose, perhaps, the investment which offers the largest expected value of the objective function, or the largest probability that the objective function will reach, at least, some critical value, or indeed some other criterion.

It is then the job of the mathematician to compute the relevant "preference functional," in most cases by first computing the distribution function of the optimum of the objective function.

This attitude, called the "wait-and-see" approach by Madansky (1960), is that which was originally called "stochastic programming" by Tintner (1955). These are not decision problems in the sense that a decision has to be made "here-and-now" about the activity levels. We wait until an observation is made on the random elements, and then solve the (deterministic) problem. Chapter I deals with such wait-and-see problems.

Chapter II contains the analysis of decision problems, in particular of so-called two-stage problems, which have been extensively studied. In these problems a decision concerning activity levels is made at once, in such a way that any emerging deviations from what would have been best, had one only known what values the stochastic elements were going to have, are in some way evaluated and affect the objective function.

In Chapter III we turn to "chance constraints," i.e. constraints which are not expected to be always satisfied, but only in a proportion of cases, or "with given probabilities." These are decision problems which reduce to problems treated in Chapter II, if the probabilities are equal to unity.

The reader should be familiar with the concepts and procedures of linear programming, and in particular with the simplex method of solving linear programs, a brief survey of which is given in Appendix I. Some concepts of nonlinear programming are also used sporadically, and for these the reader is referred to standard textbooks. All those concepts for which no special reference is given can be found, for instance, in Vajda (1961) and, more specifically, in Vajda (1967).

Features of importance for computation are mentioned, though alogorithms are not discussed in detail.

An extensive reference list is included, and in Appendix II a list is given of applications described in the literature. However, no completeness is claimed in this respect, or indeed in any other, since the subject is still being vigorously pursued by many workers.

The preparatory work, including a first draft, was carried out while the author held a Senior Research Fellowship, awarded by the Science Research Council (of Great Britain), in the Department of Mathematical Statistics of the University of Birmingham. The manuscript was completed while he was a David Brewster Cobb Senior Fellow in the Department of Transportation and Environmental Planning, in the same University.

Probabilistic Programming

I

Stochastic Programming

Parameters

Consider the problem of minimizing $c'x$, subject to $Ax \geqslant b$, $x \geqslant 0$, where c is a n-vector, b is a m-vector, and A is an m by n matrix, while x is the unknown n-vector to be determined. Let $mn+m+n = M$.

(If not mentioned otherwise, vectors are column vectors. The transpose of vectors and matrices will be indicated by a prime.)

The components can be written, parametrically, as

$$c_j = c_{j0} + c_{j1}t_{(1)} + \cdots + c_{jr}t_{(r)} \qquad (j = 1, \ldots, n)$$

$$b_i = b_{i0} + b_{i1}t_{(1)} + \cdots + b_{ir}t_{(r)} \qquad (i = 1, \ldots, m)$$

$$a_{ij} = a_{ij0} + a_{ij1}t_{(1)} + \cdots + a_{ijr}t_{(r)} \qquad (i,j \text{ as above})$$

where $r \leqslant M$, and the parameters $t_{(k)}$ $(k = 1, \ldots, r)$ have a joint probability distribution. Hence results of the theory of parametric programming can be adapted to purposes of the study of the stochastic case.

We shall assume that the support T, say, of the parameters, i.e. the smallest closed set of values $t_{(k)}$ with probability measure unity, is convex and bounded in $t = (t_{(1)}, \ldots, t_{(r)})$.

Feasibility and Convexity

We shall be interested in the relationship between the feasibility of vectors x and the parameters. It will then be possible to obtain some insight into the distribution of the minimum of the objective function $c'x$. Such knowledge will also be relevant to decision problems, since we would not, for instance, choose a value x which has no or merely a small chance of being feasible when the actual value of t emerges.

When we express all constants as linear functions of the parameters, then a linear objective function, and linear constraints, will be linear in the $t_{(k)}$, and also in the components x_j of x, separately. It follows that $T(x)$, the set of all those t for which a given $x \geqslant 0$ is feasible, is convex, and it will be polyhedral,† if T is polyhedral. The set of those t for which every $x \geqslant 0$ is feasible is the intersection of an infinity of sets $T(x)$, and hence also convex, though possibly empty.

As a simple illustration (cf. Vajda, 1970), take the case when the

† Or polytopic, but we shall continue to use the more usual term.

constraints are

$$ax \geqslant b, \qquad x \geqslant 0$$

(a case where $m = n = 1$), and a as well as b take, independently, values in the closed interval $(-1, 1)$. The (convex) sets of (a, b) for which certain chosen values of x are feasible, are shown in Figure 1.

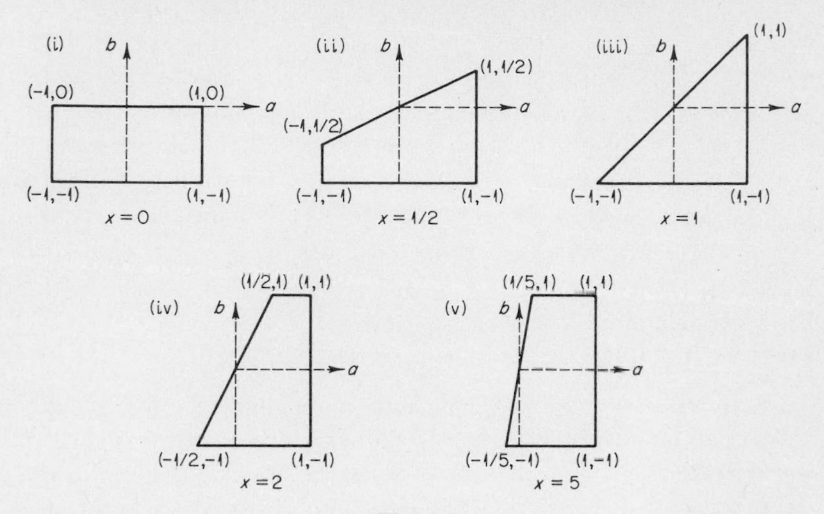

Figure 1

From the linear relationship between t and x it follows, also, that $X(t)$, the set of those $x \geqslant 0$ which are feasible for a given t is convex and polyhedral. The set of those $x \geqslant 0$ which are feasible for any t (*permanently feasible*) is the intersection of such sets and hence also convex (and possibly empty).

To take again the example above, for $a = -0.1, b = -\frac{1}{2}$, all x in [0,5] are feasible. In this example no permanently feasible x exists. If $0 < b \leqslant 1$ and $-1 \leqslant a \leqslant 0$, then $X(t) = X(a, b)$ is empty.

As far as feasibility is concerned, the objective function, and its possibly stochastic character, are irrelevant. But we must distinguish between the case when only $b' = (b_1, \ldots, b_m)$ is stochastic, and that when the elements of A (and perhaps those of b as well) are stochastic.

First, we deal with the case when only b is stochastic. Then the set of those t for which $X(t)$ is not empty is convex.

PROOF: If

$$Ax \geqslant b(t_1)$$

has a solution x_1, and

$$Ax \geqslant b(t_2)$$

has a solution x_2, then

$$Ax \geqslant b(\lambda t_1 + \mu t_2), \quad \text{with} \quad \lambda, \mu \geqslant 0, \qquad \lambda + \mu = 1$$

has also a solution, e.g. $\lambda x_1 + \mu x_2$. ■

The set of those $x \geqslant 0$ for which $T(x)$ is not empty is also convex.

PROOF: If

$$Ax_1 \geqslant b(t)$$

has a solution t_1, and

$$Ax_2 \geqslant b(t)$$

has a solution t_2, then

$$A(\lambda x_1 + \mu x_2) \geqslant b(t)$$

has also a solution, e.g. $b(\lambda t_1 + \mu t_2)$, since T is convex. ■

A theorem of Kall (1966) gives conditions for A to be such that some $x \geqslant 0$ exists whatever the values of the coefficients b, when the constraints are equations, $Ax = b$. Clearly the matrix A must then have rank m, and we assume that the first m columns, written $A_1, \ldots, A_m$, are independent. In fact, A must have more than m columns, because it is impossible that some b_0, and also $-b_0$ be represented as nonnegative linear combinations of the same m independent columns. Hence $k = n - m \geqslant 1$.

Kall's Theorem

For $Ax = b$ to have a nonnegative solution x for all b it is necessary and sufficient that there exist

$$\mu_j \geqslant 0 \qquad \text{and} \qquad \lambda_j < 0$$

such that

$$\sum_{j=m+1}^{m+k} \mu_j A_j = \sum_{j=1}^{m} \lambda_j A_j \tag{A}$$

where A_j is the jth column of the matrix A.

This condition is necessary: Let $b = \sum_{j=1}^{m} \beta_j A_j$. Such β_j exist, and for some b they will all be negative. Then, $Ax = \sum_{j=1}^{m+k} x_j A_j$ equals $\sum_{j=1}^{m} \beta_j A_j = b$, we have

$$\sum_{j=1}^{m} (\beta_j - x_j) A_j = \sum_{j=m+1}^{m+k} x_j A_j$$

and since $x_j \geqslant 0$ and $\beta_j < 0$, we can identify μ_j with x_j, and λ_j with $\beta_j - x_j$.

The condition is also sufficient: Let again

$$b = \sum_{j=1}^{m} \beta_j A_j.$$

The set of β_j is unique for a given b. If they are all nonnegative, then $Ax = b$ has a solution $x_j = \beta_j \geqslant 0$ for $j = 1, \ldots, m$, and $x_j = 0$ for $j = m+1, \ldots, m+k$.

On the other hand, if one at least of the β_j is negative, then we must find another representation of b in terms of A_j, with only nonnegative coefficients. Let, then, (A) hold. The largest of the ratios β_j/λ_j ($j \leqslant m$) is positive, if at least one of the β_j found above is negative. We may assume, without loss of generality, that the largest of these ratios is reached at $j = m$. Because $A_1, \ldots, A_m$ are independent, this is also true of $A_1, \ldots, A_{m-1}$ and $\sum_{j=1}^{m} \lambda_j A_j = A_0$, say. Thus the representation

$$b = \sum_{j=0}^{m-1} \gamma_j A_j$$

is also unique. We show, next, that all γ_j are nonnegative.

We have

$$b = \gamma_0 \sum_{j=1}^{m} \lambda_j A_j + \sum_{j=1}^{m-1} \gamma_j A_j$$

while we have, also $b = \sum_{j=1}^{m} \beta_j A_j$. Because both representations of b are unique, we must have

$$\beta_j = \gamma_j + \gamma_0 \lambda_j \qquad \text{for} \quad j = 1, \ldots, m-1$$

and

$$\beta_m = \gamma_0 \lambda_m.$$

Hence

$$\gamma_0 = \frac{\beta_m}{\lambda_m} > 0$$

and

$$\gamma_j = \beta_j - \lambda_j \gamma_0 = \lambda_j \left[\frac{\beta_j}{\lambda_j} - \frac{\beta_m}{\lambda_m} \right] \geqslant 0 \qquad \text{for} \quad j = 1, \dots, m-1.$$

To complete the argument, we remember that also $A_0 = \sum_{j=m+1}^{m+k} \mu_j A_j$ and that the μ_j are $\geqslant 0$. Thus

$$b = \sum_{j=1}^{m-1} \gamma_j A_j + \sum_{j=m+1}^{m+k} \gamma_0 \mu_j A_j$$

and in this representation all coefficients are nonnegative.

For instance, if $A = [I, -I]$, where I is the identity matrix of order m, then $\mu_j = 1$, $\lambda_j = -1$ for all j for such a set of coefficients, and indeed the set

$$x_{i1} - x_{i2} = b_i \qquad (i = 1, \dots, m)$$

has a nonnegative solution for any value of b_i.

We turn now to the case where A is stochastic. Simple examples show that then the sets of t, and x, respectively, for which $X(t)$ and $T(x)$ are not empty, are not always convex. For instance,

$$(t-3)x_1 + (1-t)x_2 \geqslant 1$$

cannot be satisfied, with nonnegative x_1 and x_2, if $1 \leqslant t \leqslant 3$. The set of those values, for which $X(t)$ is not empty, i.e. the set of values of t outside this interval, is not convex.

The set of those $x \geqslant 0$ for which $T(x)$ is not empty is not convex when the constraint is

$$(-1+2t)x_1 + (2-4t)x_2 \geqslant 1, \qquad 0 \leqslant t \leqslant 1.$$

The constraint can be written

$$(-1+2t)(x_1 - 2x_2) \geqslant 1$$

and because the first factor is in the interval $[-1, 1]$, the second must be $\geqslant 1$, or $\leqslant -1$. The set of such x_1 and x_2 is not convex.

Optimality and Convexity

We shall now consider optimality rather than feasibility. In this case we must also take into account whether or not c is stochastic. We assume again that $c'x$ is to be minimized.

The set $X^{\circ}(t)$ of those $x \geqslant 0$ which are optimal for a given t is known to be convex and polyhedral from elementary linear programming theory. Of course, it could be empty, and we start by studying the set of those t for which it is not empty.

If only c is stochastic, then this set is convex.

PROOF: If

$$c(t_1)'x_1 \leqslant c(t_1)'x$$

for all feasible x, and also

$$c(t_2)'x_2 \leqslant c(t_2)'x$$

for all feasible x, then when

$$\lambda, \mu \geqslant 0 \qquad \text{and} \qquad \lambda + \mu = 1$$

$$c(\lambda t_1 + \mu t_2)'x = \lambda c(t_1)'x + \mu c(t_2)'x$$

is bounded from below, for all feasible x, by

$$\lambda c(t_1)'x_1 + \mu c(t_2)'x_2.$$

Therefore a finite minimum exists for $t = \lambda t_1 + \mu t_2$ as well. ■

The convexity of the set of those t for which an optimum exists, when only c is stochastic, was also proved by Simons (1962), who called this set *admissible*.

If only b is stochastic, then the set of those t which admit a finite minimum is the same as the set of those t which admit a finite maximum for the dual program, and is therefore also convex, by the above argument.

Now let c as well as b be stochastic. Then the following argument† can be used to show that the set of those t for which an optimal x exists is still convex.

Let t_1 lead to a finite minimum of $c(t_1)'x$, subject to $Ax \geqslant b(t_1)$, $x \geqslant 0$. Then there is also a finite maximum to the objective function $b(t_1)'y$, subject to $A'y \leqslant c(t_1)$, $y \geqslant 0$.

Let Q be the set of those t for which $b(t)'y$, subject to $A'y \leqslant c(t_1)$, $y \geqslant 0$, has a finite maximum. It is also the set of those t which make $Ax \geqslant b(t)$ consistent, and is thus independent of t_1.

Let P be the set of those t which make the minimum of $c(t)'x$, subject to $Ax \geqslant d$, finite. It is independent of d, because it is the set of those t for which $A'y \leqslant c(t)$ is consistent. In particular, $c(t)'x$ has a finite minimum if $d = b(t_1)$ with t_1 in Q.

The set we are looking for is the intersection of P and Q, which is convex.

If A is stochastic, then the set of those t for which $X^{\circ}(t)$ is not empty is not necessarily convex.

For instance, if we have

$$(t-3)x_1 + (1-t)x_2 \geqslant 1; \qquad x_1, x_2 \geqslant 0$$

then the set $X(t)$ is not empty only when $t < 1$, or $t > 3$, as we have seen. In these regions $X^{\circ}(t)$ is not empty either when, for instance, the objective function to be minimized is $x_1 + x_2$. Its minimum is obtained for $x_1 = 0$, $x_2 = 1/(1-t)$ when $t < 1$, and for $x_1 = 1/(t-3)$, $x_2 = 0$ when $t > 3$.

We call $T^{\circ}(x)$ the set of those t for which a given $x \geqslant 0$ is optimal. Simons (1962) calls this the region of validity. He deals with the case when only c is stochastic. In this case $T^{\circ}(x)$ is convex for any x.

† From the Ph.D. Thesis in the University of Birmingham of A. S. Gonçalves (1969).

PROOF: If

$$c(t_1)x^{\circ} \leqslant c(t_1)x$$

for all feasible x, and

$$c(t_2)x^{\circ} \leqslant c(t_2)x$$

for all feasible x [and the set of feasible x is the same in both cases, because $c(t)$ does not affect feasibility], then for $\lambda+\mu=1,\ \lambda,\mu \geqslant 0$

$$c(\lambda t_1+\mu t_2)'x^{\circ} \leqslant c(\lambda t_1+\mu t_2)'x$$

for all feasible x. ■

In fact, the set $T^{\circ}(x)$ is polyhedral. If x° is not degenerate, then the corresponding shadow costs are nonpositive, and linear in t. If x° is degenerate, i.e. with basic components of value zero, then the final tableau can take different forms, and $T^{\circ}(x)$ is then the union of polyhedral sets which is itself polyhedral because, as we have just seen, that union is convex.

Now let b as well as c be stochastic. As t changes, $b(t)$ changes as well, but if x° remains optimal, then it does so within a convex set of t, and the set of $T^{\circ}(x)$ is the intersection of two convex sets, and is itself convex.

When A is stochastic, then the set $T^{\circ}(x)$ need not be convex. The following example is due to Ailsa H. Land:

Minimize

$$(7t-9)x_1+(4t-6)x_2$$

subject to

$$(5-3t)x_1+(3-2t)x_2 \leqslant 16-10t$$

$$(6-5t)x_1+(5-3t)x_2 \leqslant 22-16t$$

$$x_1, x_2 \geqslant 0.$$

Here (2, 2) is optimal for $t = 1$ (minimum -8), and also for $t = 2$ (minimum 14), but not for $t = 1.5$, when the minimum value of 2 is reached for $(\frac{4}{3}, 0)$, though (2, 2) is, of course, still feasible. (See Figure 2.)

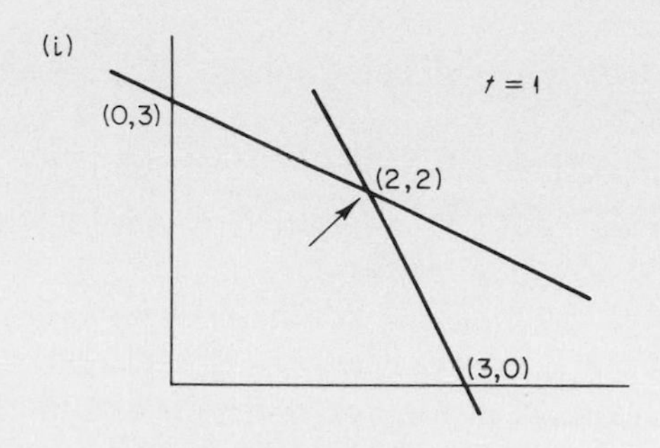

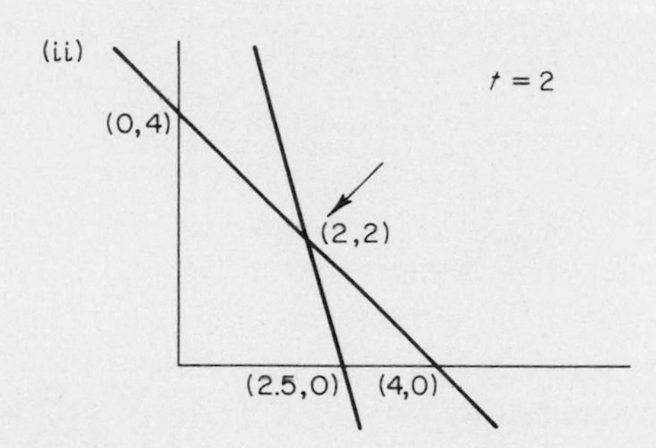

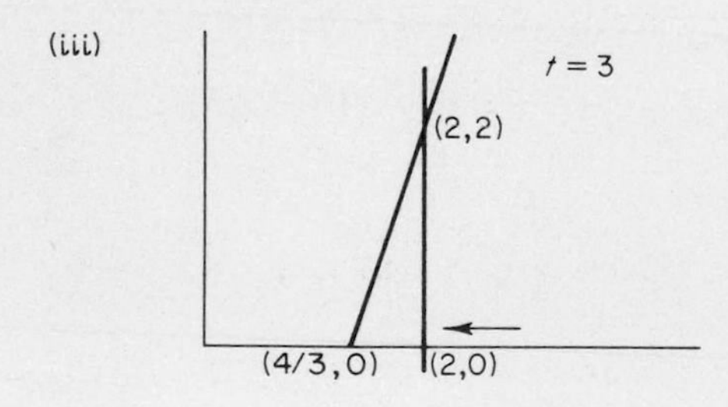

Figure 2

Decision Regions for Optimality

For the following investigation we write the constraints of a linear program $Ax = b, x \geqslant 0$, where A is an m by $n\,(> m)$ matrix with $\binom{n}{m}$ square submatrices of order m. We denote them by B_l $(l = 1, \ldots, \binom{n}{m})$ and call each of them which is not singular a basis. [Tintner (1955) called this a *selection*.] If $B_l^{-1}b$ is nonnegative, then B_l is called a feasible basis. Sengupta, Tintner, and Millham (1963) call the region of t for which B_l is feasible a *feasible* (*parameter*) *region* for B_l. Such regions, for various l, may of course overlap.

If only b is stochastic, then those t within T for which a given B_l is a feasible basis form a convex polyhedral set, because b and hence $B_l^{-1}b$ are linear in t. However, if A is stochastic, then the following is an example in which the set of t which make $B_l(t)^{-1}b(t) \geqslant 0$ need not be convex:

$$(-3+t)x_1 + (1-t)x_2 - x_3 = 1$$

$$(10-t)x_1 + (10+t)x_2 - x_4 = 10.$$

If we solve this for x_1 and x_2, then we have

$$(-40+18t)x_1 = 11t + (10+t)x_3 + (-1+t)x_4$$

$$(-40+18t)x_2 = -40 + 11t + (-10+t)x_3 + (-3+t)x_4$$

so that the basis

$$\begin{bmatrix} -3+t & 1-t \\ 10-t & 10+t \end{bmatrix}$$

is feasible for

$$40 < 18t, \qquad 11t \geqslant 0, \qquad 40 \leqslant 11t, \qquad \text{i.e. for} \quad t \geqslant \tfrac{40}{11}$$

and also for

$$40 > 18t, \qquad 11t \leqslant 0, \qquad 40 \geqslant 11t, \qquad \text{i.e. for} \quad t \leqslant 0$$

but not for

$$0 < t < \tfrac{40}{11}.$$

Similarly, we find that the pair

x_1 and x_3 is basic and feasible	when	$\frac{40}{11} \leqslant t < 10$
x_1 and x_4 is basic and feasible	when	$3 < t \leqslant \frac{40}{11}$
x_2 and x_3 is basic and feasible	when	$-10 < t \leqslant 0$
x_2 and x_4 is basic and feasible	when	$0 \leqslant t < 1$

while the pair x_3 and x_4 is never feasible.

Incidentally, we notice that all values of t are feasible except those in the interval $1 \leqslant t \leqslant 3$. This latter gap is, of course, due to the first constraint.

Bereanu (1964b, 1967) calls the set of those t which make a given feasible basis optimal for a given objective function (which may itself be stochastic) a *decision region*. If there are no degenerate solutions, then decision regions do not overlap, but two adjacent regions contain both their common boundary. When only b and/or c are stochastic, then the decision region is convex, because it is the intersection of two convex sets, viz. of the set whose elements make the basis feasible, and of the set whose elements make it dual feasible (optimal).

If A is stochastic, then the example just given has, for the basic variables, x_1 and x_2, and the objective function $2x_1 + x_2$ to be minimized, a non-convex decision region. We have then

$$2x_1 + x_2 = \frac{-40 + 33t + (10+3t)\,x_3 + (-5+3t)\,x_4}{-40 + 18t}$$

and this is optimal, and the basis is feasible, for

$$t \geqslant \tfrac{40}{11}, \qquad t \geqslant -\tfrac{10}{3}, \qquad t \geqslant \tfrac{5}{3}, \qquad \text{i.e. for} \quad t \geqslant \tfrac{40}{11}$$

and also for

$$t \leqslant 0, \qquad t \leqslant -\tfrac{10}{3}, \qquad t \leqslant \tfrac{5}{3}, \qquad \text{i.e. for} \quad t \leqslant -\tfrac{10}{3}$$

but not for $-\frac{10}{3} < t < \frac{40}{11}$. (This region is different from that found for feasibility!)

When A is stochastic, then it can also happen that isolated values of t must be excluded when considering the feasibility of a certain basis B, viz. when a value of t makes B singular. In the above example this happens at $t = 40/18$.

Approximations

Tintner (1955) uses approximations to find regions of feasibility and of optimality for a given basis and objective function. We illustrate this by a simplified example.

Minimize

$$x_1 + x_2 = C$$

subject to

$$2x_1 + x_2 - x_3 = 8 \qquad \text{and} \qquad x_1 + 2x_2 - x_4 = 7$$

$$x_1, x_2 \geqslant 0.$$

In the nonstochastic case, we get the following four basic solutions (not all of them feasible):

	x_1	x_2	x_3	x_4	C
I	3	2	0	0	5
II	7	0	6	0	7
III	4	0	0	-3	4
IV	0	$3\frac{1}{2}$	$-4\frac{1}{2}$	0	$3\frac{1}{2}$
V	0	8	0	9	8
VI	0	0	-8	-7	0

Now assume that the coefficients of x_1 in the two constraints can take any real value, independently of one another. This does not affect solutions IV, V or VI, but it does have an effect on solutions I, II and III. To ascertain the effect of a deviation d_1 of a_{11} from 2, and that of a deviation d_2 of a_{21} from 1, we difference the matrix equation $Bx = b$, and obtain

$$\delta B \cdot x + B \cdot \delta x = 0, \qquad \text{i.e.} \quad \delta x = -B^{-1} \cdot \delta B \cdot x.$$

Applying this to our example, we have for δx in case I

$$-\begin{bmatrix} 2 & 1 \\ 1 & 2 \end{bmatrix}^{-1} \begin{bmatrix} d_1 & 0 \\ d_2 & 0 \end{bmatrix} \begin{bmatrix} 3 \\ 2 \end{bmatrix} = \begin{bmatrix} -\frac{2}{3} & \frac{1}{3} \\ \frac{1}{3} & -\frac{2}{3} \end{bmatrix} \begin{bmatrix} 3d_1 \\ 3d_2 \end{bmatrix}$$

$$= \begin{bmatrix} -2d_1 + d_2 \\ d_1 - 2d_2 \end{bmatrix}$$

and in cases II and III, respectively,

$$\begin{bmatrix} -7d_2 \\ 7d_1 - 14d_2 \end{bmatrix} \quad \text{and} \quad \begin{bmatrix} -2d_1 \\ -2d_1 + 4d_2 \end{bmatrix}.$$

It follows that basis B_1 leads to

$$x_1 = 3 - 2d_1 + d_2, \qquad x_2 = 2 + d_1 - 2d_2$$

$$C_1 = 5 - d_1 - d_2$$

while B_2 and B_3 lead, respectively, to

$$x_1 = 7 - 7d_2, \qquad x_3 = 6 + 7d_1 - 14d_2$$

$$C_2 = 7 - 7d_2$$

and

$$x_1 = 4 - 2d_1, \qquad x_4 = -3 - 2d_1 + 4d_2$$

$$C_3 = 4 - 2d_1.$$

B_4 is not feasible, B_5 remains unaffected with $C_5 = 8$, and B_6 is again not feasible.

The decision region for B_1, i.e. the set of t for which C_1 is optimal, is given by those values of d_1 and d_2 which make C_1 smaller than C_2, C_3, or C_5, i.e. by

$$-d_1 + 6d_2 \leqslant 2, \qquad d_1 - d_2 \leqslant -1, \qquad -d_1 - d_2 \leqslant 3$$

and similarly we obtain the decision region for

$$C_2 \text{ by } \quad d_1 - 6d_2 \leqslant -2, \qquad 2d_1 - 7d_2 \leqslant -3, \qquad -7d_2 \leqslant 1$$

$$C_3 \text{ by } \quad -d_1 + d_2 \leqslant 1, \qquad -2d_1 + 7d_2 \leqslant 3, \qquad -2d_1 \leqslant 4$$

$$C_5 \text{ by } \quad d_1 + d_2 \leqslant -3, \qquad 7d_2 \leqslant -1, \qquad 2d_1 \leqslant -4.$$

Of course, this procedure cannot claim more than approximate validity. For instance suppose that a_{11} changes from 2 to 3, i.e. $d_1 = 1$. Then B_1 will lead, from

$$3x_1 + x_2 - x_3 = 8$$

$$x_1 + 2x_2 - x_4 = 7$$

to $x_1 = \frac{9}{5}$, $x_2 = \frac{13}{5}$, while the approximate formula above gives $x_1 = 3-2 = 1$, $x_2 = 2+1 = 3$, which values do not, in fact, satisfy the new first equation.

Inequalities

The following statements refer to the manner in which the minimum of the objective function depends on t and on b. We formulate them in a way suitable for nonlinear programs as well.

Let the constraints be $A(x) \geqslant b$, possibly including the requirement $x \geqslant 0$, where A is continuous and concave in x in the interior of the feasible region, and let the objective function $C(t, x)$ be concave in t in some closed convex region R for every feasible x, and convex in x.

1. The minimum of $C(t, x)$ is a convex function of b.

PROOF: Let $b = b_1$, and the minimum of $C(t, x)$ be $C(t, x_1) = v(b_1)$. Also, let $b = b_2$ and the minimum of $C(t, x)$, $C(t, x_2) = v(b_2)$. Moreover, let the minimum of $C(t, x)$ for $b_0 = \lambda b_1 + \mu b_2$ ($\lambda + \mu = 1$; λ, $\mu \geqslant 0$) be obtained at x_3. This x_3 may be equal to, or different from $\lambda x_1 + \mu x_2$, though the latter is a feasible vector, because from the concavity of $A(x)$ it follows that

$$A(\lambda x_1 + \mu x_2) \geqslant \lambda A(x_1) + \mu A(x_2) \geqslant b_0 .$$

Thus

$$v(b_0) = C(t, x_3) \leqslant C(t, \lambda x_1 + \mu x_2) \leqslant \lambda v(b_1) + \mu v(b_2) \qquad \blacksquare$$

If $A(x)$ is linear, i.e. Ax, then the minimum is also convex in the elements of the matrix A. The proof is analogous to that just given.

2. The minimum of $C(t, x)$ is a concave and continuous function of t in the interior of R (Mangasarian, 1964).

PROOF: Let

$$\min C(t_1, x) = C(t_1, x_1) \qquad \text{and} \qquad \min C(t_2, x) = C(t_2, x_2).$$

Then

$$\begin{aligned}\min_x C(\lambda t_1 + \mu t_2) &\geqslant \min_x [\lambda C(t_1, x) + \mu C(t_2, x)] \\ &\geqslant \min_x \lambda C(t_1, x) + \min_x \mu C(t_2, x) \\ &= \lambda \min_x C(t_1, x) + \mu \min_x C(t_2, x).\end{aligned}$$

Thus $\min_x C(t, x)$ is a concave function of t in R, and hence a continuous function of t in the interior of R. ■

Probability Distributions

So far, we have not referred to any probability distribution. If the set T forms just one single decision region, i.e. if the optimal basis is the same for all t, then the question of the distribution of $c'x$ can be approached by methods which analyze the dependence of solutions of a set of linear algebraic equations on the coefficients. This is the method of Babbar (1955); its shortcomings for linear programming have been pointed out by Wagner (1955).

If only c is stochastic, and if we denote the decision regions by S_k $(k = 1, \ldots, K)$, then for each k the optimal x is independent of t, though the optimal value of the objective function will, of course, depend on t.

Assume now that every t lies in some decision region. If for t in S_k we denote the set of t for which $c(t)' x^o \leqslant z$ by $S_k(z)$, where x^o is the optimal vector for that decision region, then the distribution function of the minimum is

$$\sum_k \int_{S_k(z)} f(t)\, dt = \sum_k H_k(z)$$

where $f(t)$ is the probability density of t, and dt is the volume element in the space of T.

In general, approximations will have to be used, as e.g. in Tintner (1955), in Sengupta, Tintner, and Millham (1963), and in Sengupta, Tintner, and Morrison (1963), using sample values.

The case of one single parameter component, either in c or in b, is considered by Bereanu (1963a). In these cases the decision regions are the intervals of parameter values found by classical methods of parametric programming (see, e.g., Gass and Saaty, 1955) and the distribution function of the optimal value is easily determined.

We discuss the following two cases in detail (cf. Bereanu, 1963a):

Case (i)	*Case (ii)*
Find	Find
$f_1(t,x) =$ minimum of $(c+td)'x$	$f_2(t, x) =$ minimum of $c'x$
subject to	subject to
$Ax = b, \qquad x \geqslant 0$	$Ax = b + td, \qquad x \geqslant 0$

where t is a random variable with a continuous and strictly increasing distribution function $\text{Prob}(t \leqslant z) = T(z)$.

We assume that the constraints are consistent in (i), and that they are consistent for all values of t in (ii). Moreover, we assume in (i) that there are no degenerate solutions and that the feasible region is bounded, and in (ii) that there are no degenerate solutions for its dual, and that for at least one t the feasible region is bounded.

Let the value which t can take be restricted to lie within the finite interval $[t_0, t_p]$. We know from the theory of parametric programming that this interval can be subdivided into decision regions, i.e. into subintervals $[t_{k-1}, t_k]$ $(k = 1, \ldots, p)$ such that for values of t within any one interval the optimal vector x does not change. We denote this vector by x^k. (At the limits of an interval, say at t_k, both the vectors x^k and x^{k+1} are optimal.) Therefore the distribution function $F_i(z)$ of the minimal value of the objective function, i.e. the probability of $f_i(t,x) \leqslant z$ $(i = 1,2)$

will be the sum $\sum_{k=1}^{p} H_k(z)$, where $H_k(z)$ is the joint probability of $t_{k-1} \leqslant t < t_k$ and of $f_i(t, x^k) \leqslant z$.

Consider now the two cases separately.

Case (*i*). We have to find $H_k(z)$ for $1 \leqslant k \leqslant p$. Consider the inequalities $t_{k-1} \leqslant t < t_k$ and $(c+td)' x^k < z$ for a given k. For t within that interval we find x^k by the usual linear programming methods.

If $d' x > 0$, then $(c+td)' x^k < z$ when $t < (z - c' x^k)/d' x^k$, so that then

$$H_k(z) = 0 \qquad \text{when} \quad \frac{z - c' x^k}{d' x^k} \leqslant t_{k-1}$$

$$H_k(z) = T(t_k) - T(t_{k-1}) \qquad \text{when} \quad \frac{z - c' x^k}{d' x^k} > t_k$$

and

$$H_k(z) = T((z - c' x^k)/d' x^k) - T(t_{k-1}) \qquad \text{otherwise.}$$

If $d' x^k < 0$, then $(c+td)' x^k < z$ when $t > (z - c' x^k)/d' x^k$, so that then

$$H_k(z) = 0 \qquad \text{when} \quad \frac{z - c' x^k}{d' x^k} > t_k$$

$$H_k(z) = T(t_k) - T(t_{k-1}) \qquad \text{when} \quad \frac{z - c' x^k}{d' x^k} \leqslant t_{k-1}$$

and

$$H_k(z) = T(t_k) - T\left(\frac{z - c' x^k}{d' x^k}\right) \qquad \text{otherwise.}$$

Finally, if $d'x^k = 0$, then $(c+td)'x^k = c'x^k$, so that then

$$H_k(z) = 0 \qquad \text{when} \quad c' x^k \geqslant z$$

and

$$H_k(z) = T(t_k) - T(t_{k-1}) \qquad \text{when} \quad c' x^k < z \; .$$

Case (*ii*). In this case x^k will consist of a term without, and one with t, thus:

$$x^k = u^k + tv^k.$$

An argument analogous to that for case (i) produces the following results:

If $c'v^k > 0$, then $c'x^k = c'(u^k + tv^k) < z$ when $t < (z - c'u^k)/c'v^k$, so that then

$$H_k(z) = 0 \qquad \text{when} \quad \frac{z - c'u^k}{c'v^k} \leqslant t_{k-1}$$

$$H_k(z) = T(t_k) - T(t_{k-1}) \qquad \text{when} \quad \frac{z - c'u^k}{c'v^k} \geqslant t_k$$

and

$$H_k(z) = T\left(\frac{z - c'u^k}{c'v^k}\right) - T(t_{k-1}) \qquad \text{otherwise.}$$

If $c'v^k < 0$, then

$$H_k(z) = 0 \qquad \text{when} \quad \frac{z - c'u^k}{c'v^k} > t_k$$

$$H_k(z) = T(t_k) - T(t_{k-1}) \qquad \text{when} \quad \frac{z - c'u^k}{c'v^k} \leqslant t_{k-1}$$

and

$$H_k(z) = T(t_k) - T\left(\frac{z - c'u^k}{c'v^k}\right) \qquad \text{otherwise.}$$

Finally, if $c'v^k = 0$, then

$$H_k(z) = 0 \qquad \text{when} \quad c'u^k \geqslant z$$

and

$$H_k(z) = T(T_k) - T(t_{k-1}) \qquad \text{when} \quad c'u^k < z.$$

If the right-hand sides of the constraints as well as the objective function depend on one (and the same) parameter, then the objective function will, for a given solution vector, be of second degree in the parameter, and the various intervals to be considered depend also on the roots of a quadratic equation. The determination of the probabilities proceeds as before: Bereanu (1963) gives the explicit formulas and solves also a case where the coefficients of the constraints are parametric. No new principles are involved.

An application of these ideas to the transportation problem is being dealt with in two communications by B. Bereanu (1963b, c).

Another case in which definite results can be stated is that of a "positive" linear program, a type discussed by Bereanu (1967).

When A, b, and c are positive matrices with probability one, then for any finite t there exists an optimal solution, except perhaps for a set of measure zero (when a submatrix of A becomes singular). This is so, because then the primal as well as the dual problem have solutions, and hence finite optimal solutions. Therefore the union of all decision regions is T.

Finally, we mention that Prékopa (1966) has given sufficient conditions for the optimal value to be asymptotically normally distributed.

The value of a matrix game in game theory can be determined by linear programming. It is the maximum of the reciprocal of $x_1 + \cdots + x_n$, subject to

$$a_{i1} x_1 + \cdots + a_{in} x_n \geqslant 1 \qquad (i = 1, \ldots, m)$$

$$x_j \geqslant 0 \qquad (j = 1, \ldots, n).$$

The distribution of the value of the game, when the a_{ij} are stochastic, is the subject of Thomas and David (1967).

II

Decision Problems

A Decision Problem

A simple stochastic decision problem is contained in an early paper by Freund (1956), who considered the cost of minimizing $C = c'x$, where the components of the vector c had a joint normal distribution with means m_j and covariance matrix V. C is then also normally distributed with mean $M = m'x$ and variance $S^2 = x'Vx$.

If all coefficients and constants of the constraints are fixed, and if we want to minimize the expected value of C, then the problem is reduced to the deterministic program of minimizing $m'x$. But let us assume that we want to minimize the expected *utility* of C, which we define as

$$1 - \exp(-aC)$$

where a is a positive constant, which economists call a measure of the *aversion to risk*.

In this case we have to minimize

$$\frac{1}{S\sqrt{2\pi}} \int_{-\infty}^{\infty} (1 - \exp(-aC)) \cdot \exp\left[\frac{-(C-m)^2}{2S^2}\right] dC$$

$$= 1 - \frac{1}{S\sqrt{2\pi}} \int_{-\infty}^{\infty} \exp\left[-aC - \frac{(C-M)^2}{2S^2}\right] dC$$

$$= 1 - \exp(-aM + \tfrac{1}{2}a^2 S^2) \cdot \frac{1}{S\sqrt{2\pi}} \int_{-\infty}^{\infty} \exp\left[\frac{-(C-M+aS^2)^2}{2S^2}\right] dC$$

$$= 1 - \exp(-aM + \tfrac{1}{2}a^2 S^2).$$

The problem is thus reduced to the deterministic program of minimizing $aM - \frac{1}{2}a^2 S^2$, a function which is quadratic in x.

The *Active* Approach

In Tintner (1960) and in Sengupta, Tintner, and Millham (1963) the authors speak of an *active* approach when the constraints $\sum_{j=1}^{n} a_{ij} x_j \leqslant b_i$ $(i = 1, \ldots, m)$ are replaced by

$$a_{ij}x_j \leqslant u_{ij}b_i \qquad (i = 1,\ldots,m;\ j = 1,\ldots,n)$$

where the u_{ij} are nonnegative values, such that $\sum_{j=1}^{n} u_{ij} = 1$, for all i. In other words, the eventual allocation of resources b_i to activities will be done according to rates u_{ij} to be chosen optimally now.

From Sengupta, Tintner, and Morrison (1963), and Sengupta (1966) one might get the impression that this active approach is *the* Here-and-Now attitude. In the former paper, and in Sengupta, Tintner, and Millham (1963) the authors compare results of the original and of the modified (active) approach when the stochastic elements have taken specific values.

The constraints of the active approach imply $\sum_j a_{ij}x_j \leqslant b_i$ for all i, but also further conditions, so that the minimum of $c'x$ in the active case cannot be smaller than that in the *passive*, wait and see case. But if all a_{ij} and all b_i are nonnegative, then for any solution $\bar{x}_1,\ldots,\bar{x}_n$ of $\sum_{j=1}^{n} a_{ij}x_j \leqslant b_i$ we can write

$$u_{ij} = \frac{a_{ij}\bar{x}_j}{\sum_{=1}^{n} a_{ij}\bar{x}_j}$$

and have then $u_{ij} \geqslant 0$, $\sum_{j=1}^{n} u_{ij} = 1$, and $a_{ij}\bar{x}_j \leqslant u_{ij}b_i$.

Hence, if we choose u_{ij} equal to

$$\frac{a_{ij}x_j^{\circ}}{\sum_{j=1}^{n} a_{ij}x_j^{\circ}}$$

in the first place, where x° is the optimal vector, then the two approaches give the same minimum. The decision variables u_{ij} are those which are also found by minimizing the objective function subject to $a_{ij}x_j \leqslant u_{ij}b_i$ (all i,j), taking the x_j and the u_{ij} to be the unknowns.

Examples:

Minimize $2x_1 - x_2$, subject to

$2x_1 + x_2 \leqslant 8$	$2x_1 \leqslant 8u_{11}, \quad x_2 \leqslant 8(1-u_{11})$
$x_1 + 2x_2 \leqslant 7$	$x_1 \leqslant 7u_{21}, \quad 2x_2 \leqslant 7(1-u_{21})$
$x_1, x_2 \geqslant 0$	$x_1, x_2 \geqslant 0, \quad 0 \leqslant u_{11}, u_{21} \leqslant 1.$

The answer turns out to be:

$x_1^\circ = 0, \quad x_2^\circ = 3\frac{1}{2}$	$x_1^\circ = 0, \quad x_2^\circ = 3\frac{1}{2}, \quad u_{11}^0 = u_{21}^0 = 0$
$2x_1^\circ - x_2^\circ = -3\frac{1}{2}$	$2x_1^\circ - x_2^\circ = -3\frac{1}{2}.$

On the other hand, if $a_{ij}x_j^\circ < 0$, for some i,j, but $\sum_{j=1}^{n} a_{ij}x_j^\circ > 0$ for that i, then the two approaches give different answers, as we see in the following example:

Maximize $2x_1 + x_2$, subject to

$x_1 + x_2 \leqslant 1$	$x_1 \leqslant u_{11}, \quad 2x_1 \leqslant u_{21}$
$2x_1 - x_2 \leqslant 1$	$2x_1 \leqslant u_{12}, \quad -x_2 \leqslant u_{22}$
$x_1, x_2 \geqslant 0$	$u_{11} + u_{12} = 1, \quad u_{21} + u_{22} = 1$
	$x_j, u_{ij} \geqslant 0.$

The final answer is now:

$x_1^\circ = \frac{2}{3}, \quad x_2^\circ = \frac{1}{3}$	$x_1^\circ = x_2^\circ = u_{11}^0 = u_{12}^0 = \frac{1}{2}$
$2x_1^\circ + x_2^\circ = \frac{5}{3}$	$u_{21}^0 = 1, \quad u_{22}^0 = 0$
	$2x_1^\circ + x_2^\circ = \frac{3}{2}.$

The paper by Sengupta, Tintner, and Morrison (1963) contains the following statement: If there is a unique solution x of the active problem for given u_{ij} when the coefficients are replaced by their means, and if the distributions of the coefficients are symmetric and mutually independent, then for variations which do not change the basis, the expected value of the objective function is not smaller than the optimal (maximal) value when the coefficients are replaced by their expectations. The same is true if we replace "objective function" by "solution vector."

This is proved for the case $n = m = 2$ and the objective function $x_1 + x_2$, to be maximized, by a binomial expansion of

$$x_i = \frac{(\bar{b}_i + \Delta b_i)\, u_{ii}}{(\bar{a}_{ii} + \Delta a_{ii})} \qquad (i = 1, 2).$$

Two-Stage Programming under Uncertainty†

Let it be required to minimize $c'x$, subject to $Ax = b$, $x \geqslant 0$ and possibly to some other constraints as well. Here c and x are n-vectors, b is an m-vector, and A is an m by n matrix.

Assume now that b is not precisely known, but only its distribution function, with finite mean Eb. It is then plainly impossible to demand that x be determined in such a way that Ax equal whatever value of b. The discrepancy between Ax and b will itself be a random variable, whose distribution function depends on x. We can then argue that we have to pay a penalty for any discrepancy, and we might decide to minimize the sum of $c'x$ and the expected value of such potential penalties.

We could make various assumptions about the penalties to be paid. A fairly general formulation, due independently to Beale (1955) and to Dantzig (1955) is as follows:

Minimize

$$c'x + E \min_{y} d'y$$

† Or "under risk." (See the Introduction.)

subject to

$$Ax + By = b$$
$$x, y \geqslant 0$$

where d and y are n_1-vectors and B is an m by n_1 matrix.

This can be interpreted to mean that a nonnegative vector x must be found "here and now," before the actual values of the components of b become known, and that, when they become known, a *recourse* y must be found from the following *second-stage program*:†

Find y which minimizes $d'y$, subject to

$$By = b - Ax, \qquad y \geqslant 0,$$

where b and x are known (and so are, of course, d, B, and A). Denote the minimum by $Q(x, b)$. The original two-stage problem can then be written as the *equivalent deterministic* program:

Minimize

$$c'x + Q(x), \qquad x \geqslant 0$$

where

$$Q(x) = EQ(x, b).$$

We shall write $Q(x, b) = \infty$ if the constraints are inconsistent, and $Q(x, b) = -\infty$ if the solutions of the constraints are unbounded below. This needs careful definition of the expected value, $Q(x)$, which we take for granted. [Compare Walkup and Wets, 1967. We might mention here that in an analogous situation of a number of decision makers in the same period, rather than of the same decision maker in a sequence of periods, Radner (1955, 1959) speaks of a *team*.]

† We shall consistently call the original problem the *two-stage problem*, and refer to the *second-stage program*.

The Complete Problem

We shall deal first with the simple case which arises when $B = [I, -I]$, where I is the identity matrix of order m, and $d' = [f', g']$, f and g being m-vectors. This case, explicitly studied by Beale (1955), was called by Wets (1966a) the complete problem, and later by Walkup and Wets (1970) a program with simple recourse (though with this definition f, g, and A might also be stochastic), and they denote by *complete recourse* a more general case which we mention later.

This simple case can be described by saying that the penalty is proportional to the absolute value of the discrepancy, in a manner defined rigorously below, but that the factor of proportionality is different in the two cases $Ax > b$ and $Ax < b$.

Let, then, the penalty for a unit of undersupply of the ith item—to which the ith constraint refers—be f_i and that for a unit of oversupply be g_i.

Formally, writing A_i for the ith row of A, we have for all i the following penalties.

$$f_i(b_i - A_i x) \quad \text{when} \quad b_i \geqslant A_i x$$

$$g_i(A_i x - b_i) \quad \text{when} \quad A_i x \geqslant b_i.$$

We introduce a variable

$$y_i^+ = b_i - A_i x \quad \text{when this is positive}$$

and

$$y_i^+ = 0 \qquad \text{otherwise}$$

and also a variable

$$y_i^- = A_i x - b_i \quad \text{when this is positive}$$

and

$$y_i^- = 0 \qquad \text{otherwise.}$$

Writing $f' = (f_1, \ldots, f_m)$ and similarly for g', y^+ and y^-, our problem can now be written:

Minimize

$$c'x + E(f'y^+ + g'y^-) \quad \text{with regard to} \quad x, y^+, \text{and } y^-$$

subject to

$$y^+ - y^- = b - Ax$$

$$x, y^+, y^- \geqslant 0.$$

Because we cannot have, at the same time, oversupply and undersupply, y_i^+ and y_i^- must not be both positive, for any i.

We assume, realistically, that all f_i as well as all g_i are nonnegative, and that they are not simultaneously zero, for any i. Then the smallest value of $E(f'y^+ + g'y^-)$ will in any case be obtained when at least one of the y^+ and y^- is zero for each i.

Beale (1961) has pointed out that for this to happen it is sufficient that $f+g > 0$, which is a weaker condition than that about f and g separately. For we may add to the objective function

$$E\{g'(y^+ - y^- - b + Ax)\} \qquad \text{or} \qquad E\{-f'(y^+ - y^- - b + Ax)\}$$

(they are both zero), and obtain, respectively,

(*i*) $c'x + E(f+g)'y^+ - g'Eb + g'Ax$, or

(*ii*) $c'x + E(f+g)'y^- + f'Eb - f'Ax$.

If $f+g > 0$, then either y^+ or y^- will be minimized, and y^+ and y^- will not both be positive.

This observation is relevant if, in the case $A_i x > b_i$ we obtain, in fact, a salvage value $h_i(ax - b_i)$, instead of paying a penalty for the oversupply. In practice h_i will be smaller than f_i, so that $f_i - h_i$, which now replaces $f_i + g_i$, will still be positive.

Adding (i) and (ii), and dividing by 2, we can write the objective function as

$$c'x + \tfrac{1}{2}E(f+g)'(y^+ + y^-) + \tfrac{1}{2}(f-g)'Eb - \tfrac{1}{2}(f-g)'Ax$$

and this is, remembering $y^+ + y^- = |b - Ax|$, and dropping the constant term, equal to

$$(c' - \tfrac{1}{2}(f' - g')A)x + \tfrac{1}{2}E(f+g)'|b - Ax|.$$

Charnes, Kirby, and Raike (1967) also obtain this expression and call it *of the constrained median type*. The constraints containing y^+ and y^- can then be ignored.

In this simple case the second-stage program reads:

Minimize

$$f'y^+ + g'y^-$$

subject to

$$y^+ - y^- = b - Ax; \qquad y^+, y^- \geqslant 0$$

where b and x are known. These constraints are never contradictory, whatever b and x are given.

The dual to this program is:

Maximize

$$\sum_{i=1}^{m} z_i(b_i - A_i x)$$

subject to

$$z_i \leqslant f_i, \; -z_i \leqslant g_i \qquad (i = 1, \ldots, m)$$

and these constraints are compatible if and only if $f+g \geqslant 0$, as we shall assume from now on. Therefore the primal of the second-stage program has a finite minimum. The original, two-stage problem has a finite minimum if also $z'A \leqslant c$. Williams (1965) has derived these results by a different method.

The dual of the second-stage program is easily solved.

When

$$b_i - A_i x < 0$$

set $z_i = -g_i$ (i.e. as small as possible), when

$$b_i - A_i x > 0$$

set $z_i = f_i$ (i.e. as large as possible), and when

$$b_i - A_i x = 0$$

then any value of z_i between the given limits will do. Thus the expected value of the optimal z_i is

$$\begin{aligned} -g_i \int_{-\infty}^{A_i x} dF_i(b_i) + f_i \int_{A_i x}^{\infty} dF_i(b_i) &= f_i - (f_i + g_i) \int_{-\infty}^{A_i x} dF_i(b_i) \\ &= \pi_i(A_i x) \end{aligned}$$

where $F_i(b_i)$ is the marginal distribution function of b_i. If $F_i(b_i^{\,0}) = 0$, and $F_i(b_i^{\,1}) = 1$ (not excluding infinite values for $b_i^{\,0}$ or $b_i^{\,1}$), then for

$$A_i x \leqslant b_i^{\,0} \qquad \text{we have} \quad \pi_i(A_i x) = f_i$$

and for

$$A_i x \geqslant b_i^{\,1} \qquad \text{we have} \quad \pi_i(A_i x) = -g_i.$$

The minimum of the second-stage program equals the maximum of its dual, and the latter is a sum of $z_i^{\circ}(b_i - A_i x)$, where z_i° is the optimal value of the dual variable.

We shall study now $Q_i(x)$, defined as follows [see (i) and (ii) above]:

$$Q_i(x) = (f_i + g_i) \int_{A_i x}^{\infty} (b_i - A_i x)\, dF(b_i) - g_i\, Eb_i + g_i A_i x$$

$$= (f_i + g_i) \int_{-\infty}^{A_i x} (A_i x - b_i)\, dF_i(b_i) + f_i\, Eb_i - f_i A_i x.$$

The integral

$$\int_{-\infty}^{y_i} (y_i - b_i)\, dF_i(b_i)$$

where we have written y_i for $A_i x$, can be transformed by integration by parts (assuming this to be legitimate) into

$$\int_{-\infty}^{y_i} F(b_i)\, db_i$$

so that

$$Q_i = f_i\, Eb_i - f_i y_i + (f_i + g_i) \int_{-\infty}^{y_i} F_i(b_i)\, db_i.$$

Williams (1965) deduced from this form that if b_i has positive density in a finite interval, then the minimum is always obtained for some x. Otherwise further conditions must hold for this to be true. More general results on this point are contained in Charnes, Kirby, and Raike (1967).

As a function of $A_i x = y_i$, $Q_i(x) = R_i(y_i)$, say, and the last formulas for Q_i show that, where $R_i(y_i)$ is differentiable—which is the case when F_i is continuous—we have $dR_i(y_i)/dy_i = -\pi_i(y_i)$. Hence for

$$A_i x \leqslant b_i^0, \qquad \frac{dR_i(y_i)}{dy_i} = -f_i$$

and for

$$A_i x \geqslant b_i^1, \qquad \frac{dR_i(y_i)}{dy_i} = g_i$$

so that in these regions $R_i(y_i)$ is linear in y_i.

It is easily established that $R_i(y_i)$ is a continuous function of y_i, and that the only points of discontinuity of the derivative are the points of discontinuity of $F_i(b_i)$. For rigorous detail, see Wets (1966a and 1966b).

If b_i^0 and b_i^1 are finite, then we can write

$$A_i x = Eb_i - h_{i1} + h_{i2} + h_{i3}$$

with

$$Eb_i - h_{i1} \leqslant b_i^0 \qquad \text{(and hence } h_{i1} \geqslant 0)$$

$$0 \leqslant h_{i2} \leqslant b_i^1 - b_i^0$$

$$0 \leqslant h_{i3}$$

and the further proviso that h_{i2} should be positive only if $Eb_i - h_{i1} = b_i^0$ (which means if $A_i x \geqslant b_i^0$) and that h_{i3} should be positive only if $h_{i2} = b_i^1 - b_i^0$ (i.e. if $A_i x \geqslant b_i^1$).

Then

$$Q_i(x) = (f_i + g_i) \int_{b_i^0}^{A_i x} (A_i x - b_i)\, dF_i(b_i) + f_i(h_{i1} - h_{i2} - h_{i3}).$$

In view of the provisos just stated, the integral has the same value as

$$\int_{b_i^0}^{b_i^0 + h_{i2}} (h_{i2} - b_i + b_i^0)\, dF_i(b_i) + h_{i3}.$$

Hence

$$Q_i(x) = (f_i + g_i) \int_{b_i^0}^{b_i^0 + h_{i2}} (h_{i2} - b_i + b_i^0)\, dF_i(b_i) - f_i h_{i2} + f_i h_{i1} + g_i h_{i3}.$$

If b_i has a uniform distribution in (b_i^0, b_i^1), then the integral in the expression for $Q_i(x)$ reduces to $h_{i2}/2(b_i^1 - b_i^0)$, and the problem of minimizing $\sum_i Q_i(x)$ subject to linear constraints is then one of quadratic programming.

For other forms of $F_i(b_i)$, some approximations are mentioned by Beale (1961) and by Wets (1966a).

Examples

We quote here an example contained in Dantzig (1955) and repeatedly referred to in later papers as well. It is as follows:

Minimize

$$x + E \min 2y$$

subject to

$$0 \leqslant x \leqslant 100$$

$$x + y \geqslant b$$

$$y \geqslant 0.$$

It can be reformulated as:

Minimize

$$x_1 + E \min(2y^+ + 0y^-)$$

subject to

$$x_1 + x_2 = 100$$

$$x_1 + y^+ - y^- = b$$

$$x_1, x_2, y^+, y^- \geqslant 0.$$

Let b be uniformly distributed in the interval (70, 80). We then have

$$Q(x) = 0.2 \int_{70}^{70+h_2} (h_2 - b + 70)\, db - 2h_2 + 2h_1$$

$$= 0.1{h_2}^2 - 2h_2 + 2h_1$$

so that we minimize

$$x_1 + 0.1{h_2}^2 - 2h_2 + 2h_1$$

subject to

$$x_1 + x_2 = 100$$

$$x_1 + h_1 - h_2 - h_3 = 75$$

$$h_1 \geqslant 5, \qquad h_2 \leqslant 10$$

$$x_1, x_2, h_1, h_2, h_3 \geqslant 0.$$

This is solved by $h_1 = 5$, $h_2 = 5$, $x_1 = 75$, $x_2 = h_3 = 0$, (this satisfies the provisos) so that the objective function has minimum value 77.5.

This approach bypasses, in effect, the second-stage program. It is of interest, though, to see other methods of computing the result.

Charnes, Cooper, and Thompson (1965a) use the formula

$$Ey^+ = E(b-x)^+ = \tfrac{1}{2}\{E|b-x| + Eb - x\}$$

so that

$$x + 2Ey^+ = E|b-x| + Eb.$$

The minimum of $E|b-x|$ is obtained at the median of b (hence the reference in the title of the paper to medians) i.e. at 75, and when b is uniformly distributed between 70 and 80, then $E|75-x| = 2.5$, so that $\min E|b-x| + Eb = 2.5 + 75 = 77.5$, as above.

In an earlier paper, Madansky (1960) proceeded as follows:

The second-stage program is to minimize αy, subject to $x+y \geqslant b$, $y \geqslant 0$. In Dantzig's example $\alpha = 2$, but we work here with (i) $\alpha = 2$, and also with (ii) $\alpha = 4$.

In view of the restricted interval of possible values of b, $x + E \min \alpha y$ equals x when $x \geqslant 80$, and it equals

(i) $x + 2Eb - 2x = 150 - x$, or

(ii) $x + 4Eb - 4x = 300 - 3x$ when $x \leqslant 70$.

When x lies between 70 and 80, then the objective function is

(i) $x + 0.2 \int_x^{80} (b-x)\, db = 77.5 + \dfrac{(x-75)^2}{10}$, or

(ii) $x + 0.4 \int_x^{80} (b-x)\, db = 78.75 + \dfrac{(x-77.5)^2}{5}$.

This is minimized in case (i) at $x = 75$, and has value 77.5, while in case (ii) it is minimized at 77.5 and has value 78.75. When x lies outside the interval, then the smallest values are larger than those just quoted. Hence the latter are the minimum values.

Similarly, we deal with the nonlinear problem (cf. Mangasarian and Rosen, 1964):

Minimize

$$x^2 + E \min 2y^2$$

subject again to

$$0 \leqslant x \leqslant 100$$

$$x + y \geqslant b$$

$$y \geqslant 0$$

where b is uniformly distributed in the interval (70, 80).

The solution of the second-stage program

$$\min 2y^2 \qquad \text{subject to} \quad x + y \geqslant b, \quad y \geqslant 0$$

is

$$y = 0 \qquad \text{when} \quad b \leqslant x$$

$$y = b - x \qquad \text{when} \quad b \geqslant x.$$

Therefore $x^2 + E \min 2y^2$ equals x^2 when $x \geqslant 80$, and it equals $x^2 + 2E(b-x)^2 = 3x^2 - 300x + 33800/3$ when $x \leqslant 70$. When x lies between 70 and 80, the objective function is

$$x^2 + \frac{1}{5}\int_x^{80} (b-x)^2 \, db = -\frac{x^3}{15} + 17x^2 - 1280x + \frac{102400}{3}.$$

The derivative of all these expressions is positive for $x \geqslant 50$, so that the minimum of the objective function is reached at $x = 50$, when it has the value 11300/3.

We have seen that the two-stage problem can be written:

Minimize

$$c'x + \sum \left\{ -f_i y_i + (f_i + g_i) \int_{-\infty}^{y_i} F_i(b_i)\, db_i \right\}$$

subject to

$$Ax - y = 0, \qquad x \geqslant 0.$$

We have omitted the irrelevant terms $f_i Eb_i$, which are independent of x and y.

$F_i(b_i)$ is a monotonic, nondecreasing function, and therefore the integral is convex in y_i. The problem is therefore one of convex programming, and the Kuhn–Tucker saddlepoint theorem applies. Williams (1965) derives from this a characterization of an optimal solution.

The *Lagrangean*

$$L(x, y; z) = (c' - z'A)x - (f - z)'y + \sum_{i=1}^{m} (f_i + g_i) \int_{-\infty}^{y_i} F_i(b_i)\, db_i$$

satisfies the inequalities

$$L(x, y; z^o) \geqslant L(x^o, y^o; z^o) \geqslant L(x^o, y^o; z)$$

for some vector z^o, every $x \geqslant 0$ and every y and z, if and only if x^o, y^o is an optimal solution of the two-stage problem.

In other words, $\min_{x, y} L(x, y; z^o) = L(x^o, y^o; z^o)$, if that minimum exists. Then $(c' - z'A) \geqslant 0$ in view of the term in x (since otherwise there would be no lower bound), and $x_i^o > 0$ only if $c_i - (z'A)_i = 0$.

As regards y, the minimum occurs when the derivative (provided it is continuous)

$$-(f_i - z_i^o) + (f_i + g_i) F_i(y_i) = 0 \qquad \text{for all } i$$

that is

$$F_i(y_i^{\circ}) = \frac{(f_i - z_i^{\circ})}{(f_i + g_i)} \qquad \text{for all those } i \text{ for which } f_i + g_i > 0.$$

If the derivative is not continuous, then the last equation must be replaced by

$$F_i(y_i^{\circ} - \varepsilon) \leqslant \frac{f_i - z_i^{\circ}}{f_i + g_i} \leqslant F_i(y_i^{\circ} + \varepsilon).$$

Williams (1966) has derived approximate formulas for linear programs, as follows:

The objective function (see above) is

$$\begin{aligned} c'x + Q(x) &= c'x + \sum_i f_i Eb_i - \sum_i f_i y_i + \sum (f_i + g_i) \int_{-\infty}^{y_i} F_i(b_i)\, db_i \\ &= \Phi(x,y) + f' Eb \end{aligned}$$

with $Ax = y$.

Take for b any fixed value b^*, such that a finite minimum exists for $c'x$, subject to $Ax = b^*$, $x \geqslant 0$, or, equivalently, a finite maximum exists for $u'b^*$, subject to $A'u \leqslant c$.

Let the respective optimal variables be $\bar{x}$ and $\bar{u}$, so that by the Duality Theorem $c'\bar{x} = \bar{u}'b^*$.

Because of $Ax = y$, we have

$$\Phi(x,y) = -\bar{u}'Ax + \bar{u}'y + \Phi(x,y).$$

If the optimal solution of the original problem is x°, $Ax^{\circ} = y^{\circ}$, then for any pair x, y with $x \geqslant 0$ we have

$$\begin{aligned} \Phi(x^{\circ}, y^{\circ}) &\geqslant \min\{-\bar{u}'Ax + \bar{u}'y + \Phi(x,y)\} \\ &= \min(c' - \bar{u}'A)x \\ &\quad + \min(\bar{u} - f)'y + \sum_i (f_i + g_i) \int_{-\infty}^{y} F_i(b_i)\, db_i \end{aligned}$$

Now $\bar{u}'A \leqslant c$, so that $\min_{x \geqslant 0} (c' - \bar{u}'A)x = 0$.

The terms in y are convex, and the minimum is attained when $y_i = \hat{y}_i$ such that $F_i(\hat{y}_i) = (f_i - \bar{u}_i)/(f_i + g_i)$ for all i.

Hence

$$\Phi(x^o, y^o) \geqslant \sum_i (f_i + g_i) \int_{-\infty}^{\hat{y}_i} F_i(b_i)\, db_i + \sum_i (\bar{u}_i - f_i)\, \hat{y}_i.$$

On the other hand

$$\Phi(\bar{x}, b^*) = c'\bar{x} - \sum_i f_i b_i^* + \sum_i (f_i + g_i) \int_{-\infty}^{b_i^*} F_i(b_i)\, db_i$$

and, replacing $c'\bar{x}$ by $\bar{u}'b^*$ and subtracting, we obtain

$$\Phi(\bar{x}, b^*) - \Phi(x^o, y^o) \leqslant (\bar{u} - f)'(b^* - \hat{y}) + \sum_i (f_i + g_i) \int_{\hat{y}_i}^{b_i^*} F_i(b_i)\, db_i.$$

This gives an upper bound for the error committed when we take $b = b^*$ and find $\Phi(\bar{x}, b^*)$ rather than $\Phi(x^o, y^o)$.

To illustrate, take the linear example (ii), and let $b^* = 75$ (the mean value of b). The two dual problems are:

Minimize		Maximize
x		$-100u_1 + 75u_2$
	subject to	
$-x \geqslant -100$		$-u_1 + u_2 \leqslant 1$
$x \geqslant 75$		no sign-restriction.

These are solved by

$\bar{x} = 75$	$\bar{u}_2 = 1.$

Then $F(\hat{y}) = (4-1)/4 = \frac{3}{4}$, hence $\hat{y} = 77.5$.

$$\begin{aligned} c'\bar{x} + Q(\bar{x}) &= \Phi(\bar{x}, 75) + 4Eb \\ &= 75 - 4 \times 75 + 4 \int_{70}^{75} F(b)\, db + 4 \times 75 \\ &= 80. \end{aligned}$$

The error, by taking 80 as the value of $c'x + Q(x)$ instead of the correct value, (which we have seen was 78.75) is not larger than

$$\begin{aligned} (1-4)(75-77.5) + 4 \int_{77.5}^{75} F(b)\, db &= 7.5 - 76.25 + 70 \\ &= 1.25. \end{aligned}$$

In fact, the difference is precisely that.

Discrete Values of b_i

If the b_i can only take discrete values, say b_{ik} $(k = 1, \ldots, K)$, with probabilities p_{ik}, then the constraints are

$$A_i x + y_{ik}^{+} - y_{ik}^{-} = b_{ik} \qquad (k = 1, \ldots, K)$$

$$y_{ik}^{+}, y_{ik}^{-} \geqslant 0$$

and the objective function is

$$c'x + \sum_{i=1}^{m} \{f_i \sum_k p_{ik} y_{ik}^{+} + g_i \sum_k p_{ik} y_{ik}^{-}\}$$

so that we have now a (possibly very large) linear program.

An alternative formulation is as follows:

Write

$$A_i x = u_{i0} + u_{i1} + \cdots + u_{iK} \qquad (i = 1, \ldots, m)$$

$$0 \leqslant u_{ik} \leqslant b_{ik+1} - b_{ik} \qquad (k = 1, \ldots, K-1)$$

where $b_{i0} = 0$.

With these constraints, we minimize

$$c'x + \sum_i f_i [p_{i1}(b_{i1} - u_{i0}) + p_{i2}(b_{i2} - u_{i0} - u_{i1}) + \cdots$$
$$+ p_{iK}(b_{iK} - u_{i0} - \cdots - u_{iK-1})]$$
$$+ \sum_i g_i [p_{i1}(u_{i1} + \cdots + u_{iK}) + p_{i2}(u_{i2} + \cdots + u_{iK}) + \cdots + p_{iK} u_{iK}]$$
$$= c'x + \sum_i f_i [Eb_i - u_{i0} - \cdots - u_{iK}]$$
$$+ \sum_i (f_i + g_i)[p_{i1} u_{i1} + (p_{i1} + p_{i2}) u_{i2} + \cdots$$
$$+ (p_{i1} + \cdots + p_{iK-1}) u_{iK-1}].$$

We must also ensure that u_{ik} is positive only if all u_{it} with $t < k$ have reached their upper limit. But this is certain, in view of the form of the objective function. For instance, it would not pay to increase u_{it}, with the coefficient $-f_i + (f_i + g_i)(p_{i1} + \cdots + p_{it})$, at the expense of u_{is}, if $s < t$. [Remember that $(f+g)$ is nonnegative.]

Elmaghraby (1959), has established this equivalence by a slightly different argument. He was considering a case where penalties arose through the x_i not having, in the final answer, an anticipated value. A similar method was also used by Szwarc (1964), as applied to the Transportation Problem.

If an oversupply can not possibly be tolerated, then the constraints reduce to $A_i x + y_i = b_{i1}, x, y \geqslant 0$ and the objective function reduces to

$$c'x + \sum_i f_i [p_{i1} y_i + p_{i2}(y_i + b_{i2} - b_{i1}) + \cdots + p_{iK}(y_i + b_{iK} - b_{i1})]$$

which equals

$$c'x + \sum_i f_i y_i$$

plus a constant which is irrelevant for minimization.

The General Case, *b* Stochastic

We turn now to the more general case of two-stage problems under uncertainty, but still assuming that only b is stochastic:

Minimize

$$c'x + E \min d'y$$

subject to

$$Ax + By = b$$

$$x, y \geqslant 0.$$

Its second-stage program reads:

Find

$$Q(x, b), \quad \text{the minimum of } d'y$$

subject to

$$By = b - Ax, y \geqslant 0 \qquad \text{with given } x \text{ and } b.$$

If B is a matrix of positive elements, then it induces the condition $Ax \leqslant b$.

If B is a nonsingular square matrix, then the (unique) y of the second-stage program equals $B^{-1}(b-Ax)$, and y will be nonnegative if $B^{-1}b \geqslant B^{-1}Ax$. This will be so for all b if $\min B^{-1}b \geqslant B^{-1}Ax$.

More generally, we can write for the constraints of the two-stage problem

$$0 \leqslant x \leqslant A^*(b-By) + (I-A^*A)\,Y$$

where A^* is the generalized inverse of A, I is an identity matrix, and Y is arbitrary. For more detail of this approach see Charnes, Cooper, and Thompson (1965a).

Feasibility

Through the choice of x, the second-stage program might have contradictory constraints.

We call a vector x feasible, if it satisfies all constraints which do not depend on y (if any), and is such that the second-stage program has a solution $y \geqslant 0$, whatever the value of b. The region of feasible vectors x will be denoted by K. Of course, if $B=[I, -I]$ (the complete problem above), then $By = b-Ax, y \geqslant 0$ has a solution for any x and b, so that then K equals the intersection of the entire positive orthant, with the region defined by constraints independent of y (if any).

It is thus natural to ask which conditions on B make $By = b, y \geqslant 0$ solvable for all t. We have seen in Chapter I that for this to be the case it is necessary and sufficient that

$$\sum_{j=m+1}^{m+k} \mu_j B_j = \sum_{j=1}^{m} \lambda_j B_j$$

(where B_j is the jth column of B) have a solution with all $\mu_j \geqslant 0$ and all $\lambda_j < 0$. When $B=[I, -I]$, then $\mu_j = 1$, $\lambda_j = -1$ form such a set.

If this condition is not satisfied, then K will be smaller, but it will, in any case, be convex. This can be seen as follows:

Denote the region of feasible x for given b by K_b. This is obviously closed and convex. K is the intersection of all K_b (and possibly of the convex region defined by other constraints), and is hence also convex.

As a matter of fact, K_b is not merely convex, but polyhedral and so is K, as will now be shown.

Consider the region L_b in m-space, of all vectors $z = b - By$ with $y \geqslant 0$. This is a cone spanned by the columns of $-B$ and then translated so that its vertex is b.

Next, consider L, the intersection of all L_b. Of course, the intersection of an infinite number of polyhedra is not necessarily polyhedral. But in the present case the various L_b are parallel to one another, and from this our statement follows. A proof making use of the concept of polar cones is given in Wets (1966c).

We show now that, as a consequence of L being polyhedral, the same is true of K. Since L is a convex polyhedron, it can be defined by the constraints $Vz \geqslant h$. K consists of all those vectors x for which $Ax = b - By$ has a solution $y \geqslant 0$ for any b, i.e. of all those for which $Ax = z$ with some z in L. This is the same as to say that it consists of all those vectors x for which $VAx \geqslant h$. Hence it is a polyhedron.

We are now interested in finding sufficient and necessary conditions for a given nonnegative x to be feasible. First, let x and b be given. Then from the lemma of Farkas $By = b - Ax$ has a solution in nonnegative y if (and trivially only if) all solutions u of $u'B \geqslant 0$ are also solutions of $u'(b - Ax) \geqslant 0$. In other words, it is necessary and sufficient for x to be in K_b, that the minimum of $u'(b - Ax)$, subject to $u'B \geqslant 0$ be nonnegative. For x to be in K, the condition must hold for all possible vectors b.

Further progress can be made on these lines if the original constraints were given as inequalities, thus

$$Ax + By \leqslant b \qquad \text{or} \qquad Ax + By + z = b$$

$$x, y, z \geqslant 0.$$

Then we have to add, to $u'B \geqslant 0$, the further constraint $u \geqslant 0$. If b has a lower bound, $b_{\min}$, then we have also (if $u \geqslant 0$)

$$u'(b - Ax) \geqslant u'(b_{\min} - Ax)$$

so that if min $u'(b_{\min} - Ax)$, subject to $u'B \geqslant 0, u \geqslant 0$ is nonnegative, then x will be in K.

If $b_{\min}$ is itself a possible vector b, then this condition is clearly also necessary. However, the following example (from Wets, 1966b) shows that otherwise this is not always true:

Let

$$A = \begin{bmatrix} 1 & 0 \\ 0 & 1 \end{bmatrix}, \qquad B = \begin{bmatrix} 1 & -1 \\ 1 & 1 \end{bmatrix}$$

and let b take values in the region defined by $-1 \leqslant b_1 \leqslant 0, b_2 \leqslant 2$, $b_1 + b_2 \geqslant 0$, i.e. in the quadrilateral with vertices $(0,0), (0,2), (-1,2)$, $(-1,1)$. Then $b_{\min} = (-1,0)$, which is not a possible vector.

In this case the vector $\bar{x} = (0,0)$ is in K, because

$$0 + y_1 - y_2 \leqslant b_1, \qquad 0 + y_1 + y_2 \leqslant b_2$$

has a nonnegative solution (y_1, y_2) for all (b_1, b_2), for instance $(0, b_2)$. However, the test problem

$$\min u'(b_{\min} - A\bar{x}) = \min[u_1(-1-0) + u_2(0-0)] = \min(-u_1)$$

subject to

$$u_1 + u_2 \geqslant 0$$

$$-u_1 + u_2 \geqslant 0$$

$$u_1, u_2 \geqslant 0$$

does not give a nonnegative optimum. In fact, $-u_1$ is unbounded from below.

If $\bar{u}$ is an optimal solution of the test problem, then x, to be feasible, must be such that $\bar{u}(b_{\min}-Ax) \geqslant 0$. We could add this (deterministic) constraint to those of the problem, thereby cutting off a portion of K.

Optimality

We study now the minimum of the objective function $d'y$ of the second-stage program, whose constraints are

$$By = b - Ax, \qquad y \geqslant 0$$

with given x and b. This minimum is convex in x.

PROOF: Let $\min_y d'y$ equal $v(x_1)$ when $x = x_1$, and $v(x_2)$ when $x = x_2$, and let it be reached, respectively, when $y = y_i$, $i = 1, 2$. Then $x_0 = \lambda x_1 + \mu x_2$ $(\lambda + \mu = 1; \lambda, \mu \geqslant 0)$ is feasible, because K is convex, and when $x = x_0$, then the minimum $v(x_0)$ of $d'y$ might, or might not be obtained at $y_0 = \lambda y_1 + \mu y_2$, but cannot be larger than $d'y_0$. Hence

$$v(x_0) \leqslant f'y_0 = \lambda v(x_1) + \mu v(x_2). \qquad \blacksquare$$

Because $E \min_y d'y = Q(x)$ is a positively weighted combination of the several minima, convexity in x follows for that expected value as well, and hence also for $c'x + Q(x)$. The essence of this proof is contained, for instance, in Beale (1955).

We proceed now to the derivation of a *necessary* condition for a vector $\bar{x}$ to be optimal.

The dual of the second-stage program reads:

Maximize

$$z'(b - Ax)$$

subject to

$$B'z \leqslant d$$

where b and x are known.

Denote the optimal vector z by $\bar{z}(b,x)$. From the duality theorem of linear programming we have

$$\bar{z}(b,x)'(b-Ax) = \min d'y$$

when

$$Ax + By = b \qquad \text{and} \qquad y \geqslant 0.$$

Consequently, if $\bar{x}$ is optimal for the original two-stage problem, then

$$c'\bar{x} + E\bar{z}(b,\bar{x})'(b-A\bar{x}) \leqslant c'x + E\bar{z}(b,x)'(b-Ax)$$

for all feasible x. Moreover

$$\bar{z}(b,x)'(b-A\bar{x}) \leqslant \bar{z}(b,\bar{x})'(b-A\bar{x})$$

because when $x = \bar{x}$, then it is $z = \bar{z}(b,\bar{x})$ which maximizes $z'(b-Ax)$. Take expectations with regard to b, and adding $c'\bar{x}$, we have

$$c'\bar{x} + E\bar{z}(b,x)'(b-A\bar{x}) \leqslant c'\bar{x} + E\bar{z}(b,\bar{x})'(b-A\bar{x})$$

and hence, with the previous inequality,

$$c'\bar{x} + E\bar{z}(b,x)'(b-A\bar{x}) \leqslant c'x + E\bar{z}(b,x)'(b-Ax).$$

Dropping $E\bar{z}(b, x)'b$, we obtain the required necessary condition

$$[c' - E\bar{z}(b,x)'A]\,\bar{x} \leqslant [c' - E\bar{z}(b,x)'A]\,x.$$

We have here followed Dantzig and Madansky (1961). In that paper the following *sufficient* condition has also been derived (Theorem 4):

If $\bar{x}$ is feasible and if

$$[c' - E\bar{z}(b,\bar{x})'A]\,\bar{x} \leqslant [c' - E\bar{z}(b,\bar{x})'A]\,x$$

for all feasible x, then $\bar{x}$ is optimal. [Note the difference between $\bar{z}(b,x)$ and $\bar{z}(b,\bar{x})$ which appear, respectively, in the two statements.]

We prove this last theorem for the case when there is only a finite number of possible vectors b, say $b_1,\ldots,b_K$. In this case the constraints are

$$Ax + By_k = b_k \qquad (k = 1,\ldots,K)$$

and we have to minimize

$$c'x + \sum_k p_k d' y_k \qquad \text{with} \quad x, y_1, \ldots, y_K \geqslant 0.$$

[The dual of the decomposition algorithm of Dantzig and Wolfe (1960) is a convenient way of solving such a problem.]

If $\bar{x}, \bar{y}_k$ satisfy the constraints of the two-stage problem, then this set is optimal if and only if we can find vectors $z_1,\ldots,z_K$ satisfying the constraints of the dual

$$A'z_1 + \cdots + A'z_K \leqslant c$$

$$B'z_k \leqslant p_k \cdot d$$

and such that (they satisfy complementary slackness, viz.)

$$(c - A'z_1 - \cdots - A'z_K)\,\bar{x} = 0$$

$$(p_k d' - B'z_k)\,\bar{y}_k = 0 \qquad (k = 1,\ldots,K).$$

Now the conditions involving B are satisfied by the optimal solution of the dual kth second-stage programs when $\bar{y}_k$ is the optimal solution of its primal. Thus we see that $\bar{x}$ is optimal for the two stage problem if and only if

$$c' - Ez'A \geqslant 0 \qquad \text{and} \qquad [c' - Ez'A]\bar{x} = 0$$

where the z are the optimal solutions of the dual of the second-stage program, given $\bar{x}$.

While in the finite case the converse of this theorem is also easily established, Dantzig and Madansky (1961) mention that it is unknown whether the converse of the sufficiency theorem is true or not in the general case.

Similar theorems are found in Madansky (1963). A necessary and sufficient condition for optimality is also given in Elmaghraby (1960).

The optimal value of the objective function $d'y$ will be finite, if the dual has a feasible solution, i.e. when $B'z \leqslant d$ is not contradictory. Since this condition is independent of b and of x, the same holds for $E \min d'y$ in the two-stage problem. In the Appendix to this chapter we mention a necessary condition for $B'z \leqslant d$ to have a solution z for given d, when B satisfies the condition given in Kall's Theorem (in Chapter I) which make $By = b - Ax$ feasible for any right-hand side.

The General Case, A and b Stochastic

We generalize now by allowing also the elements of A and b to be random, while still keeping B fixed. This case is called *of fixed recourse* in Walkup and Wets (1970). We keep c also fixed, or think of it as of the mean (assumed finite) of a random vector.

An example of this case is the active approach of Tintner (1960) and of others mentioned earlier in this Chapter. In the notation used there, it is u_{ij} which has to be determined "here and now," and x_j in the second-stage program; therefore we write Y_j for x_j, and X_{ij} for u_{ij}, to conform to our present notation. Thus we transcribe

$$-b_i u_{ij} + a_{ij} x_j \leqslant 0 \qquad \text{into} \qquad AX + BY = 0$$

where A is the diagonal matrix with diagonal entries $-b_1, \ldots, -b_m$ repeated n times, X' is the vector $(X_{11}, \ldots, X_{m1}, X_{12}, \ldots, X_{m2}, \ldots, X_{mn})$, B is a block diagonal matrix with the jth block being

$$\begin{matrix} a_{1j} & 1 & 0 & \cdots & 0 \\ a_{2j} & 0 & 1 & \cdots & 0 \\ & & \vdots & & \\ a_{mj} & 0 & 0 & \cdots & 1 \end{matrix}$$

(The a_{ij}, and hence B, are not stochastic here.) Y' is the vector $(Y_1, z_{11}, \ldots, z_{m1}, Y_2, z_{12}, \ldots, z_{m2}, Y_n, z_{1n}, \ldots, z_{mn})$. The z_{ij} are slack variables.

For the second-stage program (which is independent of d) we keep x, A, and b fixed. Denote the set of feasible $x \geqslant 0$ for given A and b by K_{Ab} [the *elementary feasibility set* of Walkup and Wets (1967)]. It is obviously closed and convex. Moreover, it is polyhedral, whenever the closure of the convex hull of the support of A and b is itself polyhedral (e.g. when the elements can take only discrete values), or whenever A and b are independently distributed and the closure of the convex hull of the support of A is polyhedral (Walkup and Wets, 1967, Propositions 3.15 and 3.16). We do not prove these propositions here.

As before, the optimum of the second-stage program $Q(x, A, b)$, will be finite, if $B'z \leqslant d$ is not contradictory.

The objective function $c'x + Q(x)$ of the two-stage problem is convex in x. This is proved in the same way as in the case when only b was stochastic.

Walkup and Wets (1969) have also proved that $Q(x)$ is either unbounded from below for all x in K_{Ab}, or finite and has finite variance. It is lower semicontinuous provided it is not unbounded for any x. However, the following example shows that it need not be continuous.

Minimize

$$E \min_y y_1$$

subject to

$$\begin{aligned} a_1 x_1 - y_1 - y_2 \qquad &= 0 \\ a_2 x_2 \qquad - y_2 - y_3 &= 0 \\ x, y &\geqslant 0 \end{aligned}$$

where (a_1, a_2) can take the values $(2^{2n}, 2^{2n} - 2^n)$, $n = 1, 2, \ldots$ with probability 2^{-n}. All $x \geqslant 0$ are feasible.

The second-stage program requires to

minimize

$$y_1(n)$$

subject to

$$y_i(n) \geqslant 0 \qquad (i = 1, 2, 3)$$

$$y_1(n) + y_2(n) = 2^{2n} x_1$$

$$y_2(n) + y_3(n) = (2^{2n} - 2^n) x_2$$

where x_1, x_2, and n are given.

The solution (omitting the argument n) is

$$y_1 = 0, \qquad y_2 = 2^{2n} x_1, \qquad y_3 = (2^{2n} - 2^n) x_2 - 2^{2n} x_1$$

if

$$\frac{x_1}{x_2} \leqslant 1 - 2^{-n}$$

and

$$y_1 = (-2^{2n} + 2^n) x_2 + 2^{2n} x_1, \qquad y_2 = (2^{2n} - 2^n) x_2, \qquad y_3 = 0$$

if

$$\frac{x_1}{x_2} \geqslant 1 - 2^{-n}.$$

The expectation of the minimum, i.e. $Q(x)$, depends on x_1/x_2.

When $x_1/x_2 \leqslant 1-2^{-1}$, then it is also $\leqslant 1-2^{-n}$ for all $n \geqslant 1$, so that $\min y_1 = 0$, and $Q(x) = 0$.

When $1-2^{-n} \leqslant x_1/x_2 \leqslant 1-2^{-(n+1)}$, then $\min y_1 = 2^{2n}(x_1-x_2)+2^n x_2$ as long as $x_2/(x_2-x_1) \geqslant 2^n$ and 0 afterwards, so that the expected value $Q(x)$ is again finite. However, when $x_1/x_2 \geqslant 1$, then $Q(x) = \sum_{n=1}^{\infty} [2^n(x_1-x_2)+x_2]$ diverges, unless $x_1 = x_2 = 0$. Thus in every neighborhood of $(0,0)$ there is an arbitrarily large value of $Q(x)$. This function is not continuous in that point, though it is everywhere convex.

Generalizations of the theorem stating a necessary condition for optimality are also given in Walkup and Wets (1967).

The General Case, *b*, *A*, and *B* Stochastic

Finally, we allow the elements of B to be also random. Here the simplest case is that where B is square, and nonsingular with probability 1. (Program with *stable recourse*.)

When we define now the feasibility of an $x \geqslant 0$, we must distinguish between cases where the second-stage program for that x is

always feasible (strong feasibility)

and where it is

feasible with probability 1 (weak feasibility).

This is the case called *of complete recourse* in Walkup and Wets (1970).

The difference between these two definitions is illustrated by the following example, due to Walkup and Wets (1970).

Minimize

$$x + \min E(y)$$

subject to

$$x + wy = 1$$

$$x, y \geqslant 0$$

where x, y, and w are one-component vectors and w is distributed in $[0,1]$ with zero probability for $w = 0$. For instance, the distribution function of w might be (i) $F(w) = w$, or (ii) $F(w) = w^2$.

In such a case, weak feasibility holds for $0 \leqslant x \leqslant 1$, but strong feasibility only for $x = 1$.

In case (i), we have $Ey = (1-x) \int_0^1 dw/w$, which is divergent, but in case (ii), $Ey = (1-x) \int_0^1 2\,dw = 2(1-x)$.

Walkup and Wets (1967) have established a sufficient condition for strong and weak feasibility to coincide. This is so if both the set of those t for which $B'y = t$ has a solution $y \geqslant 0$, and the set for which $B'y \leqslant t$ has any solution (without sign-restriction on y), are continuous in the elements of B.

This condition was not satisfied in the example above. There B reduces to w, $wy = t$ has a nonnegative solution y for all $t \geqslant 0$ when $w > 0$, but only for $t = 0$ when $w = 0$.

The objective function of the second-stage program is convex in x, in A and b, and concave in d.

One could consider recourses, or compensations, which are more general than those defined by $By = b - Ax$, for instance, the minimization of $c'x + d(b-Ax) + |D(b-Ax)|$ where D is a matrix. Dempster (1968) mentions that, in his unpublished doctoral dissertation, he has found sufficient conditions for this program to have a finite solution. He considers also the minimization of

$$c'x + d(b-Ax) + E(b-Ax)'H(b-Ax)$$

where H is a symmetric, nonnegative definite matrix. In this case the problem reduces to one of quadratic programming, if the constraints are linear.

If the distribution function is unknown, but it is known that it belongs to a given family, then the minimax approach can be applied, by choosing that x which minimizes the largest possible expectation of the sum of the objective function and of the penalties. Such an approach, which allows us to choose a distribution of x rather than a fixed value, and deals with general penalty functions, is studied by Iosifescu and Theodorescu (1963) and by J. Žáčková (1966). These authors use concepts and results of the theory of games.

Inequalities

A device, which suggests itself naturally when the parameters of a program are stochastic, is the replacement of the parameters by some judiciously chosen fixed values, such as, for instance, their mean values or some pessimistic values to produce a "fat solution" (Madansky, 1962), so as to be "on the safe side." We shall now try to obtain some idea of the bias, if any, which is thereby introduced into the results.

We shall, in particular, prove the following string of inequalities:

$$E\,C(b,\bar{x}(Eb)) \geqslant \min_x E\,C(b,x) \geqslant E\min_x C(b,x) \geqslant \min_x C(Eb,x)$$

where $C(b,x)$ is the minimum, with regard to y, of the (not necessarily linear) objective function $c(x)+d(y)$, subject to $A(x)+B(y) \geqslant b$, with given b and x, and $\bar{x}(Eb)$ is that vector x which minimizes $C(Eb,x)$. The operator E stands for the expectation with regard to b, a random variable.

We shall assume that $A(x)$ and $B(y)$ are concave and continuous in their respective arguments, and for the last inequality also that $c(x)$ and $d(y)$ are convex and continuous in their respective arguments. If it is required that x or y be nonnegative, then we imagine this to be included in the constraints as written.

In our previous notation, we have in the linear case

$$C(b,x) = c'x + Q(x,b),$$

so that $E\,C(b,x) = c'x + Q(x)$, to be minimized in the two-stage problem.

The first inequality holds, because $\bar{x}(Eb,x)$ is (assumed to be) a feasible x, but not necessarily that which minimizes $E\,C(b,x)$.

For the second inequality, let $\min_x E\,C(b,x)$ be obtained when $x = \bar{x}$, i.e. $\min_x E\,C(b,x) = E\,C(b,\bar{x})$, and let $\min_x C(b,x) = C(b,\bar{x}(b))$ for given b, so that $E \min_x C(b,x) = E\; C(b,\bar{x}(b))$. Now for every $b, C(b,\bar{x}(b)) \leqslant C(b,\bar{x})$, and hence

$$E \min_x C(b,x) = E\,C(b,\bar{x}(b)) \leqslant E\,C(b,\bar{x}) = \min_x E\,C(b,x). \qquad \blacksquare$$

It will be noticed that these two inequalities remain true if not only b, but also A and c are random.

For the last inequality, we refer to the theorem of Jensen (1906) which states that for a convex, continuous function $f(b)$ of a random variable b, we have

$$Ef(b) \geqslant f(Eb)$$

and this holds in the present case, because (as we now show) $\min_x C(b,x)$ is (a) convex and (b) continuous in b.

Of course, if $\min_x C(b,x)$ is linear in b, then the two sides of the last inequality have equal values.

(a) $\min_x C(b,x)$ is a convex function of b.

Let

$$\min_x C(b_1,x) = C(b_1,x_1) = c(x_1) + f(y_1)$$

and

$$\min_x C(b_2,x) = C(b_2,x_2) = c(x_2) + f(y_2)$$

where

$$x_1, x_2, y_1, y_2 \geqslant 0$$

and

$$A(x_1) + B(y_1) \geqslant b_1, \qquad A(x_2) + B(y_2) \geqslant b_2.$$

Because A and B are concave, we have for $x_0 = \lambda x_1 + \mu x_2, y_0 = \lambda y_1 + \mu y_2$, and $b_0 = \lambda b_1 + \mu b_2$ $(\lambda + \mu = 1, \lambda, \mu \geqslant 0)$

$$A(x_0) \geqslant \lambda A(x_1) + \mu A(x_2), \qquad B(y_0) \geqslant \lambda B(y_1) + \mu B(y_2)$$

and hence

$$A(x_0) + B(y_0) \geqslant \lambda[A(x_1) + B(y_1)] + \mu[A(x_2) + B(y_2)]$$

$$\geqslant \lambda b_1 + \mu b_2 = b_0.$$

Therefore $C(b_0, x_0)$, i.e. the minimum with regard to y of

$$c(x_0) + f(y), \qquad \text{subject to} \quad A(x_0) + B(y) \geqslant b_0$$

cannot exceed

$$c(x_0) + f(y_0) \leqslant \lambda[c(x_1) + f(y_1)] + \mu[c(x_2) + f(y_2)]$$

$$= \lambda C(b_1, x_1) + \mu C(b_2, x_2).$$

Consequently

$$\min_x C(b_0, x) \leqslant C(b_0, x_0)$$

$$\leqslant \lambda \min_x C(b_1, x) + \mu \min_x C(b_2, x). \qquad \blacksquare$$

An alternative proof (Madansky, 1960) for the linear case is as follows.

The dual problem to minimizing $c'x+d'y$, subject to $Ax+By \geqslant b$ reads:

Maximize

$$b'z$$

subject to

$$z'A \leqslant c, \qquad z'B \leqslant d, \qquad z \geqslant 0.$$

Let the optimal z be $\bar{z}$. Then, by the Duality Theorem

$$b'\bar{z} = C(b, \bar{x}(b)) \qquad \text{where} \quad \bar{x}(b) \text{ is the optimal } x \text{ for given } b.$$

If $\bar{z}_i$ maximizes $b_i{}'z\,(i = 1, 2)$, then

$$\begin{aligned} \min_x C(b, x) &= C(b, \bar{x}(b)) \\ &= b'\bar{z} \\ &\leqslant \lambda b'_1 \bar{z}_1 + \mu b'_2 \bar{z}_2 \\ &= \lambda C(b_1, \bar{x}(b_1)) + \mu C(b_2, \bar{x}(b_2)) \end{aligned}$$

as above.

(b) $\min_x C(b, x)$ is a continuous function of b, provided $c(x)$ and $d(y)$ are convex and continuous, and $A(x)$ and $B(y)$ are concave and continuous functions of their respective arguments.

We have seen that the set of vectors b for which there exists a feasible x is convex; hence that set of vectors b for which there exists an x minimizing $C(b, x)$ is also convex. Thus $\min_x C(b, x)$ is a convex function of b over a convex set of b, and therefore continuous at every interior point of this set.

The fact that it is also continuous at every boundary point, follows from the continuity of c, d, A and B.

These proofs are taken from Mangasarian and Rosen (1964). The linear analogue of the inequalities appeared in Madansky (1960).

In the linear context, the last inequality was also proved by Vajda (1958), as follows:

Consider the problem of minimizing

$$c'x + f'y$$

subject to

$$Ax + By \geqslant b.$$

Its dual is the problem of maximizing

$$v'b$$

subject to

$$A'v = c, \quad B'v = f, \qquad v \geqslant 0.$$

For any given b,

$$\begin{aligned} \min_x C(b,x) &= \min_{x,y} c'x + f'y \\ &= \max_v v'b \\ &= \bar{v}'(b) \cdot b. \end{aligned}$$

Consider, also, the problem of minimizing

$$c'x + f'y$$

subject to

$$Ax + By \geqslant Eb$$

whose dual requires to maximize

$$v'Eb$$

subject to

$$A'v = c, \qquad B'v = f, \qquad v \geqslant 0.$$

Again by duality,

$$\min_x C(Eb, x) = \max_x v'Eb = v_0'Eb.$$

But v_0 is feasible in the former problem as well (the constraints are in fact identical), so that

$$\max_v v'b \geqslant v_0'b \qquad \text{for every} \quad b.$$

Taking expectations on both sides, we obtain the result.

Generalizations of these results to abstract vector spaces are found in Bui Trong Lieu (1964).

The value that interests us in the two stage problem is $\min_x E\,C(b, x)$, and the inequalities provide upper and lower bounds for it.

If the range of possible values of the components of b is finite, say

$$b_{\min} \leqslant b \leqslant b_{\max}$$

then an upper bound for $\min_x E\,C(b, x)$ is $\min_x C(b_{\max}, x)$.

PROOF: For any b and x, a vector y which satisfies $A(x) + B(y) \geqslant b_{\max}$ will also satisfy $A(x) + B(y) \geqslant b$, so that $C(b_{\max}, x) \geqslant C(b, x)$ for all b. The right-hand side can therefore be replaced by $E\,C(b, x)$, and the result

$$\min_x C(b_{\max}, x) \geqslant \min_x E\,C(b, x)$$

then follows at once. The left-hand side of this inequality is called the "fat" solution in Madansky (1962).

[Another upper bound for $E \min_x C(b,x)$ has been derived by Madansky (1959).]

Let $\min_x C(Eb,x) = C(Eb,\bar{x}(Eb))$ and $\min_x E\ C(b,x) = E\ C(b,\bar{x})$ as before. Then, provided $E\,C(b,x)$ is differentiable at $\bar{x}(Eb)$, the following inequality holds:

$$\min_x E\,C(b,x) \geqslant E\,C(b,\bar{x}(Eb)) + [\bar{x} - \bar{x}(Eb)]'\ \nabla\ E\,C(b,\bar{x}(Eb))$$

where ∇ is a column vector of the partial differential operators $\partial/\partial x_i$. This follows from the fact that $E\,C(b,x)$ is convex in x, (as we have seen), and from a well-known property of convex functions. [$\bar{x}(Eb)$ could, of course, be replaced by any other point, but $\bar{x}(Eb)$ is easily determined.]

Comparing the inequality just stated, and the first of the string of inequalities proved above, it follows that equality holds if $\bar{x} = \bar{x}(Eb)$, or if $\nabla EC(b,\bar{x}(Eb))$ vanishes; e.g. if $EC(b,\bar{x}(Eb))$ is of the form $x_i^2 - 2\bar{x}_i(Eb) + k$ for each x_i.

If $\bar{x}$ defined above is independent of b, or if $C(b,x) = C(b,\bar{x})$ except on a set of b of probability measure zero, then trivially $\min_x E\,C(b,x) = E \min_x C(b,x)$. This is also true if $C(b,x)$ is linear in b. This follows from $\min_x E\,C(b,x) = \min_x C(Eb,x)$ and the second and third in the string of inequalities proved above. In this case all three expressions in that string of inequalities are of equal value.

Finally, we mention that if the set of possible vectors b is finite, then $E \min_x C(b,x) = \min_x C(Eb,x)$ if and only if $\min_x C(b,x)$ is linear in b.

We illustrate the sequence of inequalities by using examples which illustrated the solution of the two-stage problem:

Minimize

	(cf. Mangasarian and Rosen, 1964)
$x + E \min 2y$	$x^2 + E \min 2y^2$

subject to

$$0 \leqslant x \leqslant 100$$

$$x + y \geqslant b$$

$$y \geqslant 0$$

where b is uniformly distributed in the interval

$$(70, 80), \qquad \text{so that} \quad Eb = 75.$$

(i) $E\,C(b,\,\bar{x}(Eb))$ is obtained when

$\bar{x}(75) = 75, \qquad y = 0$	$\bar{x}(75) = 50, \qquad y = 25$

(See Figure 3.)

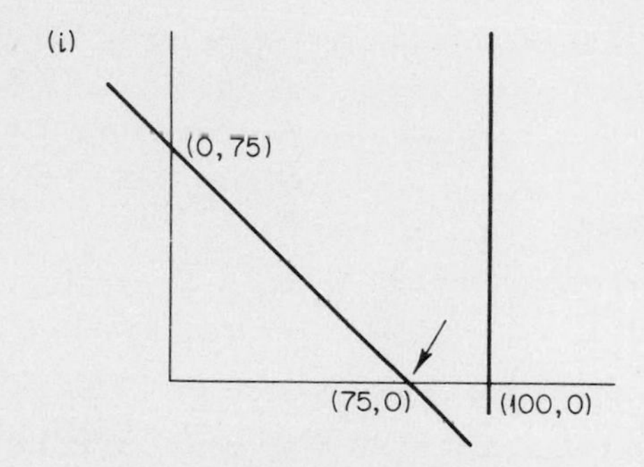

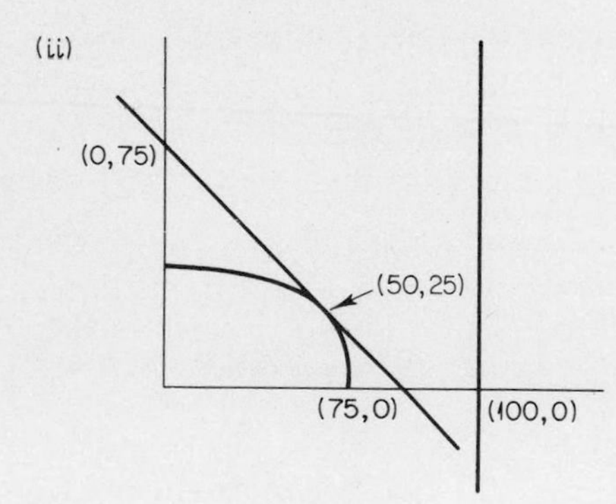

Figure 3

$$C(b, 75) = \min_y (75 + 2y)$$

subject to

$$75 + y \geqslant b$$

is solved by

$$y = 0 \qquad \text{when} \quad b \leqslant 75$$

$$y = b - 75 \qquad \text{when} \quad b \geqslant 75.$$

Hence

$$E\,C(b, 75)$$

$$= 75 + \frac{1}{10}\int_{75}^{80} (2b - 150)\,db$$

$$= 77.5.$$

$$C(b, 50) = \min_y (2500 + 2y^2)$$

subject to

$$50 + y \geqslant b$$

is solved by

$$y = 0 \qquad \text{when} \quad b \leqslant 50$$

$$y = b - 50 \qquad \text{when} \quad b \geqslant 50.$$

Hence

$$E\,C(b, 50)$$

$$= 2500 + \frac{1}{10}\int_{70}^{80} 2(b - 50)^2\,db$$

$$= \frac{11300}{3}$$

$$\sim 3767.$$

(ii) $\min_x E\,C(b, x)$

This we have already worked out in the examples above:

77.5 (at $\bar{x} = 75$).

$\frac{11300}{3}$ (at $\bar{x} = 50$).

(iii) $E \min_x C(b,x)$

$$\min_x C(b,x) = 0 \quad \text{if } b \leqslant 0$$
$$= 2b - 100 \quad \text{if } b \geqslant 100$$
and
$$= b \quad \text{if } 0 \leqslant b \leqslant 100$$
hence
$$E \min_x C(b,x) = \frac{1}{10}\int_{70}^{80} b\, db$$
$$= 75.$$

$$\min_x C(b,x) = 0 \quad \text{if } b \leqslant 0$$
$$= 10000 + 2(b-100)^2 \quad \text{if } b \geqslant 150$$
and
$$= \frac{2b^2}{3} \quad \text{if } 0 \leqslant b \leqslant 150$$
hence
$$E \min_x C(b,x) = \frac{1}{10}\int_{70}^{80} \frac{2}{3} b^2\, db$$
$$= \frac{33800}{9}$$
$$\sim 3756.$$

(iv) $\min_x C(Eb,x)$

$$C(75,x) = x \quad \text{if } x \geqslant 75$$
and
$$= 150 - x \quad \text{if } x \leqslant 75$$
hence
$$\min_x C(Eb,x) = 75.$$

$$C(75,x) = x^2 \quad \text{if } x \geqslant 75$$
and
$$= 3x^2 - 300x + 11250 \quad \text{if } x \leqslant 75$$
hence
$$\min_x C(Eb,x) = C(Eb,50)$$
$$= 3750.$$

The inequalities are satisfied by these values. We have, also

$$\min_x C(b_{\text{max}}, x) = \min_x C(80, x) \quad \text{equal to 80} \quad \text{(larger than 77.5)}$$

in the linear, and equal to 12800/3 (larger than 3767) in the nonlinear example. We see also that in both examples

$$\min_x E\, C(b, x) = E\, C(b, \bar{x}(Eb))$$

as it must be, because $\bar{x} = \bar{x}(Eb)$.

It will also be noticed that $\bar{x}$ equals the x which minimizes $C(Eb, x)$ [though the values of the respective objective functions of (ii) and (iv) differ].

This will always be so if $C(b, x)$ is a quadratic function of x and b, because the quadratic term in b is irrelevant to minimization with regard to x. This was pointed out in Simon (1956) and Theil (1957) in the context of "Dynamic Programming." See also the discussion in Madansky (1960) and his reference on page 200 to Reiter (1957).

It should be understood that the inequalities remain valid if the constraints are equations, thus $A(x) + B(y) = b$.

We might be interested in the comparison of the

$$\min E\, C(b, x) = \min_{x,y} (c'x + f'y)$$

subject to

$$Ax + By = b; \qquad x, y \geqslant 0$$

and $\min_x c'x$, subject to $Ax = Eb$.

We show by an example, that the first minimum might be larger, or it might be smaller, than the second. We choose for this purpose a slight modification of the example we have already studied in some detail:

Minimize

$$x_1 + E \min_y \alpha (y_1 + y_2)$$

subject to

$$x_1 + x_2 = 100$$

$$x_1 + y_1 - y_2 = b$$

$$x_1, x_2, y_1, y_2 \geqslant 0.$$

The second-stage program:

Minimize

$$\alpha(y_1 + y_2)$$

subject to

$$x_1 + y_1 - y_2 = b \qquad (x_1 \text{ given})$$

$$y_1, y_2 \geqslant 0$$

is solved by

$$y_1 = b - x_1, \qquad y_2 = 0 \qquad \text{when} \quad x_1 \leqslant b$$

and by

$$y_1 = 0, \qquad y_2 = x_1 - b \qquad \text{when} \quad x_1 \geqslant b.$$

Hence

$$\min x_1 + E \min_y \alpha(y_1 + y_2)$$

equals

$$x_1 + E\alpha(x_1 - b) = x_1 + \alpha(x_1 - 75) \qquad \text{when} \quad x_1 \geqslant 80$$

$$x_1 + E\alpha(b - x_1) = x_1 + \alpha(75 - x_1) \qquad \text{when} \quad x_1 \leqslant 70$$

and

$$x_1 + 0.1\alpha \int_{70}^{x_1} (x_1 - b)\, db + 0.1\alpha \int_{x_1}^{80} (b - x_1)\, db$$

$$= x_1 + \alpha\left[\frac{(x_1 - 75)^2}{10} + 2.5\right]$$

when x_1 is between 70 and 80.

It is easily computed that, for instance, when $\alpha = 2$, the minimum is found at $x_1 = 72.5$ and has value 78.75. When $\alpha = 1$, it is obtained at all values $x_1 \leqslant 70$ and has value 75. When $\alpha = 0.5$, then it is obtained at $x_1 = 0$, and has value 37.5. The first value is larger than, and the last is smaller than 75, which is the solution of the following:

$$\text{Minimize } x_1 \qquad \text{subject to} \quad x_1 + x_2 = 100, \quad x_1 = 75.$$

A similar example with diagrams is contained in Vajda (1958).

Turning now to the case of random coefficients in the objective function $f(c, x)$, we prove the following string of inequalities:

$$E \max_x f(c, x) \geqslant \max_x Ef(c, x) \geqslant Ef(c, \bar{x}(Ec)) \geqslant \max_x f(Ec, x).$$

E means now expectation with regard to c.

For the last inequality we assume that c takes values in a convex region, and also that $f(c, x)$ is convex and continuous in c for all x in the feasible region.

PROOF: Let

$$\max_x f(c, x) = f(c, \bar{x}(c))$$

and

$$\max_x Ef(c, x) = Ef(c, \bar{x}).$$

Take expectations on both sides of the first of these equations. Since

$$f(c, \bar{x}(c)) \geqslant f(c, \bar{x}) \qquad \text{for all} \quad c$$

we have

$$Ef(c, \bar{x}(c)) \geqslant Ef(c, \bar{x})$$

and this is the first inequality.

The second inequality is obvious, since $\bar{x}(Ec)$ might or might not maximize $Ef(c, x)$.

Because $f(c, \bar{x}(Ec))$ is convex and continuous in c, it follows from Jensen's inequality (1906), that

$$Ef(c, \bar{x}(Ec)) \geqslant f(Ec, \bar{x}(Ec)) = \max_x f(Ec, x). \qquad \blacksquare$$

If f is linear in c, then the second and third inequalities above become equations.

It should be noted that the sequence of these inequalities is different from that of the earlier string of inequalities, which we have proved above.

Mangasarian (1964) uses the following example as an illustration:

Maximize

$$c_1 x_1^2 + c_2^2 x_2^2 = f(c, x)$$

subject to

$$0 \leqslant x_1 \leqslant 10, \qquad 0 \leqslant x_2 \leqslant 5$$

where both c_1 and c_2 are uniformly distributed, within the ranges $-\frac{1}{2} \leqslant c_1 \leqslant \frac{1}{2}$, $0 \leqslant c_2 \leqslant 1$, so that $Ec_1 = 0$, $Ec_2 = \frac{1}{2}$.

We have then

(i) $\max_x f(c,x) = \max_x (c_1 x_1^{\,2} + c_2^{\,2} x_2^{\,2}) = 100c_1 + 25c_2^{\,2}$

when $c_1 \geqslant 0$

and

$$\max_x f(c,x) = \max_x (c_1 x_1^{\,2} + c_2^{\,2} x_2^{\,2}) = 25c_2^{\,2} \quad \text{when } c_1 \leqslant 0$$

so that

$$E \max_x (c_1 x_1^{\,2} + c_2^{\,2} x_2^{\,2}) = 12.5 + 25 \int_0^1 c_2^{\,2}\, dc$$

$$= 20\tfrac{5}{6}.$$

(ii) $E f(c,x) = \dfrac{x_2^{\,2}}{3}$

so that

$$\max_x E f(c,x) = 8\tfrac{1}{3}.$$

(iii) $\max_x (Ec_1 x_1^{\,2} + (Ec_2)^2 x_2^{\,2}) = \max_x \dfrac{x_2^{\,2}}{4}$

so that

$$\bar{x}_2(Ec) = 5$$

and

$$f(c, \bar{x}(Ec)) = c_1 x_1^{\,2} + 25c_2^{\,2}, \qquad E f(c, \bar{x}(Ec)) = 8\tfrac{1}{3}.$$

(iv) $f(Ec,x) = \dfrac{x_2^{\,2}}{4}$

so that

$$\max_x f(Ec,x) = 6\tfrac{1}{4}.$$

Appendix

It is required that the inequality $B'z \leqslant d$ have some solution z for a given d(a $m+k$-vector), where B is an m by $m+k$ matrix which satisfies the conditions on the matrix A in Kall's Theorem in Chapter I.

Theorem (Kall, 1966). For $B'z \leqslant d$ to have a solution z for given d, it is necessary that

$$\sum_{i=1}^{m} \lambda_i d_i \leqslant \sum_{i=m+1}^{m+k} \mu_i d_i$$

where λ_i and μ_i are a solution of the necessary and sufficient conditions of Kall's Theorem, applied to B instead of A.

PROOF: Let a solution of the inequality $B'z \leqslant d$ be z_0, so that $B_i' z_0 \leqslant d_i$ $(i = 1, \ldots, m+k)$, where B_i is the ith column of B. Then

$$\sum_{i=m+1}^{m+k} \mu_i B_i' z_0 \leqslant \sum_{i=m+1}^{m+k} \mu_i d_i.$$

But the left-hand side is also equal to $\sum_{i=1}^{m} \lambda_i B_i' z_0$, and this is not smaller than $\sum_{i=1}^{m} \lambda_i d_i$, because of $B_i' z_0 \leqslant d_i$, and $\lambda_i < 0$.

Hence

$$\sum_{i=1}^{m} \lambda_i d_i \leqslant \sum_{i=m+1}^{m+k} \mu_i d_i$$

follows as a necessary consequence.

When $k = 1$, then $\mu + 1$ cannot be zero and the condition can be written

$$\sum_{i=1}^{m} \lambda_i d_i \leqslant d_{m+1}$$

where the $\lambda_i < 0$ satisfy $B_{m+1} = \sum_{i=1}^{m} \lambda_i B_i$. This condition is (in this case $k = 1$) also sufficient:

Let $B_i' z = d_i (i = 1, .., m)$ have the (unique) solution z_0. We have then

$$B'_{m+1} z_0 = \sum_{i=1}^{m} \lambda_i B_i' z_0 = \sum_{i=1}^{m} \lambda_i d_i \leqslant d_{m+1}.$$

Thus z_0 is a solution of $B'z \leqslant d$.

However, for $k > 1$ the condition is not sufficient, as the following example shows:

Let

$$B = \begin{bmatrix} 1 & 1 & -2 & -1 \\ -1 & 1 & 1 & -2 \end{bmatrix}, \qquad d' = (1, 1, 1, -2)$$

and let

$$\lambda_1 = \lambda_2 = -1, \qquad \mu_3 = \mu_4 = 1.$$

Then the condition $-1-1 \leqslant 1-2$ is satisfied.

Nevertheless, the inequalities $B'z \leqslant d$, i.e.

$$z_1 - z_2 \leqslant 1, \qquad z_1 + z_2 \leqslant 1$$

$$-z_1 + 2z_2 \leqslant 1, \qquad -z_1 - 2z_2 \leqslant -2$$

are contradictory (the second and third when $z_2 \geqslant 1$, and the second and fourth when $z_2 < 1$, so that no z_2 can satisfy all four of them).

The region of those z which satisfy $B'z \leqslant d$ is bounded. Otherwise there would exist a solution $z_0 + \Delta$ to every z_0 such that $B'\Delta < 0$. But this is impossible, because then, if $Q(x) < \infty$, by Kall's first theorem we should have

$$0 \geqslant \sum_{i=m+1}^{m+k} \mu_i B_i \Delta = \sum_{i=1}^{m} \lambda_i B_i \Delta > 0$$

which is a contradiction. ■

III

Chance Constraints

A. Charnes and W. W. Cooper (1960, 1962, 1963) have initiated a probabilistic approach to programming which is different from those which we have so far considered.

In their approach, it is not required that the constraints should always hold, but we shall be satisfied if they hold in a given proportion of cases or, to put it differently, if they hold with given probabilities.

Thus we might want to minimize $c'x$, subject to $x \geqslant 0$, and to

$$\text{Prob}\,(\sum_j a_{ij}x_j \geqslant b_i) \geqslant \alpha_i \qquad (i = 1, \ldots, m)$$

where the α_i are given probabilities not equal to zero, and the symbol Prob ( ) means *the probability of* ( ).

Alternatively, we might want the joint probability that all inequalities

$$\sum a_{ij}x_j \geqslant b_i$$

hold, to be at least α. This condition is less restrictive than the previous one, if $\alpha = \prod \alpha_i$.

Because of the probabilistic aspect of the problem, it is possible to approach its solution from two different points of view.

I. We might want to find such values of the components of x, that the probability that they will satisfy the inequalities within the chance-constraints, when the values taken by the random variables become known, is at least that required. Rules for finding such values are called *zero-order rules* by Charnes and Cooper (1963, p. 24).

II. We might want to find a chance mechanism for determining values of the components of x which satisfy the inequalities within the chance constraints with at least the required probabilities. In this case the objective function would be the expected value of $c'x$.

For case I, with which we start, our main tool will consist in finding deterministic constraints (i.e. constraints not containing any probabilistic element), which are equivalent to the chance-constraints.

As a simple introductory example of this procedure [cf. Wessels (1967), who uses this example for a different purpose], consider the constraint

$$\text{Prob}\,(a(x_1 - x_2) \geqslant \tfrac{1}{2}) \geqslant \tfrac{1}{4}$$

where a is uniformly distributed between -1 and 1.

We draw a diagram of the curve $a = \frac{1}{2}(x_1 - x_2)^{-1}$ against $(x_1 - x_2)$, and indicate where $a(x_1 - x_2) \geqslant \frac{1}{2}$, i.e.

$$a \geqslant \tfrac{1}{2}(x_1 - x_2)^{-1} \qquad \text{when} \quad x_1 - x_2 > 0$$

and

$$a \leqslant \tfrac{1}{2}(x_1 - x_2)^{-1} \qquad \text{when} \quad x_1 - x_2 < 0$$

and where a lies between -1 and 1. (See Figure 4.)

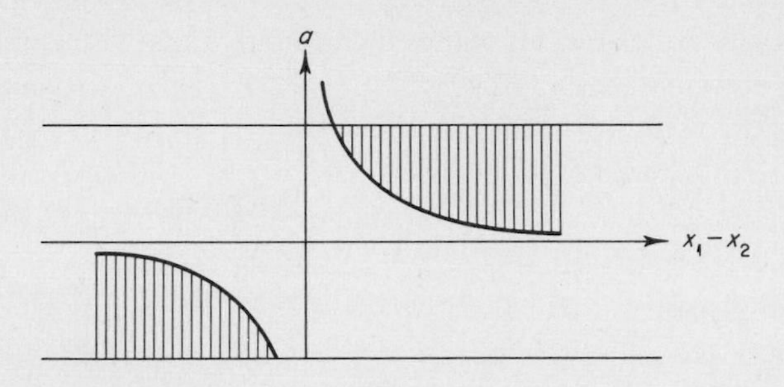

Figure 4

The required probability is at least $\frac{1}{4}$ when the relevant portion of the verticals have a length of at least $\frac{1}{2}$ [because the interval $(-1, 1)$ has a length of 2]. This is the case when $|x_1 - x_2| \geqslant 1$, and this condition is the deterministic equivalent we are looking for.

It is possible that for some $(b_1, \ldots, b_m)$ no x satisfies all $\sum_j a_{ij} x_j \geqslant b_i$. However, all the chance constraints might still be simultaneously feasible for some x, even if there is some set $(b_1, \ldots, b_m)$ such that $\sum_j a_{ij} x_j \geqslant b_i$ is not feasible for any i. This point is made by Charnes, Kirby, and Raike (to appear), and we illustrate it by the following example:

$$\text{Prob}\,(-x_1 - x_2 \geqslant b_1) \geqslant \tfrac{1}{2}$$

$$\text{Prob}\,(x_1 - x_2 \geqslant b_2) \geqslant \tfrac{1}{2}$$

with x further subject to

$$x_1 - x_2 \leqslant 1 \qquad \text{and} \qquad x_1, x_2 \geqslant 0.$$

If both b_1 and b_2 are independently, and uniformly distributed in $(-2,2)$, then both $-x_1-x_2 \geqslant b_1$, and $x_1-x_2 \geqslant b_2$ are impossible when $(b_1, b_2) = (1, 2)$. Yet, both chance constraints (and also the further constraints) are satisfied when $x_1 = x_2 = 0$.

Nevertheless, the set of chance constraints might be infeasible for some b. Charnes, Kirby, and Raike (to appear) suggest the use of artificial variables within the brackets of the chance constraints, with appropriately modified objective functions. If and only if the original constraints have no solution, then the artificial variables will appear in the optimal solution of the modified problem.

Other possibilities of interpreting and dealing with not feasible chance constraints are discussed in Charnes and Kirby (1966).

Quantile Rules

In what follows, we shall frequently apply a rule which gives a deterministic equivalent to a chance-constraint using quantiles of distribution functions.

Consider the constraint

$$\operatorname{Prob}(ax \geqslant b) \geqslant \alpha$$

where b has a known distribution function $\operatorname{Prob}(b \leqslant z) = F_b(z)$. We find the smallest value B_α such that $F_b(B_\alpha) = \alpha$, which can also be written

$$B_\alpha = F_b^{-1}(\alpha).$$

Then the constraint above is equivalent to the nonprobabilistic constraint $ax \geqslant B_\alpha$.

This is so, because if $ax \geqslant B_\alpha$, then ax will not be smaller (but might be larger) than any of those b which are not larger than B_α, and the probability of such b arising is α. On the other hand, if ax is smaller than B_α, then the probability of $ax \geqslant b$ being true is less than α. (See Figure 5.)

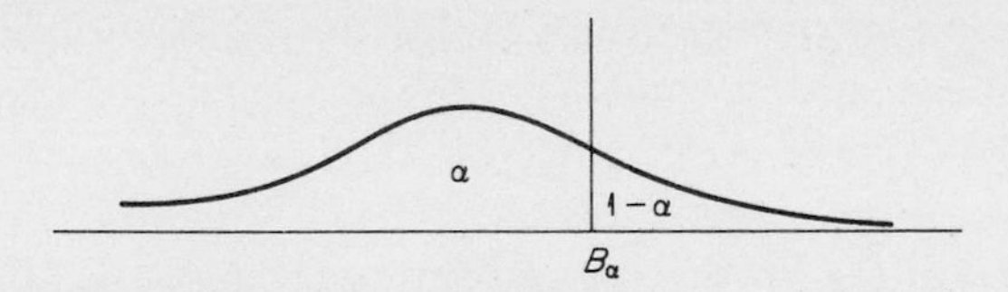

Figure 5

By an analogous argument we find the following equivalences:

$$\text{Prob}\,(ax \leqslant b) \leqslant \alpha \quad \text{is equivalent to} \quad ax \geqslant B_{1-\alpha}$$

$$\text{Prob}\,(ax \geqslant b) \leqslant \alpha \quad \text{is equivalent to} \quad ax \leqslant B_{\alpha}$$

$$\text{Prob}\,(ax \leqslant b) \geqslant \alpha \quad \text{is equivalent to} \quad ax \leqslant B_{1-\alpha}.$$

Now consider once more the constraint $\text{Prob}\,(ax \geqslant b) \geqslant \alpha$, but this time let a be the random element, with distribution function $\text{Prob}\,(a \leqslant z) = F_a(z)$.

Define a_0 to be the largest value of a for which

$$F_a(a_0) = 1-\alpha, \qquad \text{i.e.} \quad a_0 = F_a^{-1}(1-\alpha).$$

Then for $x \geqslant 0$,

$$\text{Prob}\,(ax \geqslant a_0 x) \geqslant \text{Prob}\,(a \geqslant a_0) = \alpha$$

so that

$$\text{Prob}\,(ax \geqslant b) \geqslant \alpha \quad \text{is equivalent to} \quad a_0 x \geqslant b.$$

(The inequality between the two probabilities holds if $x = 0$.)

We consider now in more detail a constraint

$$\sum_j a_{ij} x_j \geqslant b_i$$

in the following two situations:

(i) The b_i are independently distributed random variables;

(ii) the a_{ij} are random variables whose distributions have means and standard deviations, and for each i, a_{ij} is distributed independently of a_{ik} $(j \neq k)$.

Let the constraints be

$$\text{Prob}\,(\sum_j a_{ij} x_j \geqslant b_i) \geqslant \alpha_i \qquad (i = 1, \ldots, m)$$

and assume that the $\sum_j a_{ij} x_j \geqslant b_i$ are feasible for all sets of b_i with positive probability density, taking into account any other conditions on x (which could, incidentally, also be written as chance constraints, with the respective $\alpha_i = 1$).

We can deal with just one constraint at a time and drop the subscript i.

(i) Let the distribution function of b be $F_b(z)$. Then the constraint $\text{Prob}\,(\sum_j a_j x_j \geqslant b) \geqslant \alpha$ is equivalent to $\sum_j a_j x_j \geqslant F_b^{-1}(\alpha)$.

(ii) Let $\bar{a}_j$ be the mean value of a_j, and s_j its standard deviation. Then, given x, the mean value of $\sum_j a_j x_j$ is $\sum_j \bar{a}_j x_j = M$, and its standard deviation is the nonnegative value of $(\sum s_j^2 x_j^2)^{1/2} = S$ (assuming M and S exist).

We now look for the largest constant τ_N such that

$$\text{Prob}\left(\frac{\sum a_j x_j - M}{S} \geqslant \tau_N\right) = \alpha.$$

If b is not larger than $M + \tau_N S$, then it is not larger than any of those $\sum_j a_j x_j$ which are not smaller than $M + \tau_N S$, and the probability of such $\sum a_j x_j$ is α. Hence the constraint

$$\text{Prob}(\sum_j a_j x_j \geqslant b) \geqslant \alpha$$

is equivalent to the (nonstochastic) constraint

$$M + \tau_N S \geqslant b.$$

Of course, M and S contain the unknown x_j. The value of τ_N will also, in general, depend on the x_j. Therefore our approach is only reasonable—if mean and standard deviations exist and are finite and—if τ_N is, in fact, independent of x. In other words, we can proceed in the manner described if the distribution of

$$\frac{\sum_j a_j x_j - \sum_j \bar{a}_j x_j}{(\sum_j s_j^2 x_j^2)^{1/2}}$$

is the same as that of each $(a_j - \bar{a}_j)/s_j$.

This is, in fact, the case when the distributions of the a_j are normal, even if they have different means and standard deviations. It is natural to ask whether other distributions could be similarly treated.

We shall see that this is so, even though other parameters than means and standard deviations might have to be used. To investigate this point, we use characteristic functions.

The characteristic function of $\sum_j a_j x_j$ is the function

$$\prod_j E \exp(ita_j x_j)$$

where E stands for *expected value*.

For instance, if the distribution of a_j is normal, with density

$$\frac{1}{\sigma_j \sqrt{2\pi}} \exp\left[-\frac{1}{2}\left(\frac{a_j - \bar{a}_j}{\sigma_j}\right)^2\right]$$

then

$$E[\exp(ita_j x_j)] = \exp(i\bar{a}_j x_j t - \tfrac{1}{2}\sigma_j^2 x_j^2 t^2)$$

and the characteristic function of $\sum_j a_j x_j$ is

$$\exp[it \sum_j \bar{a}_j x_j - \tfrac{1}{2}t^2 \sum_j \sigma_j^{\,2} x_j^{\,2}]$$

which is the characteristic function of a normal distribution with mean $\sum_j \bar{a}_j x_j = M$, and standard deviation $(\sum_j \sigma_j^{\,2} x_j^{\,2})^{1/2} = S$. Therefore, if we choose τ_N such that

$$\frac{1}{\sqrt{2\pi}} \int_{-\infty}^{\tau_N} \exp(-\tfrac{1}{2}x^2)\, dx = 1 - \alpha$$

then

$$\text{Prob}\left(\frac{\sum_j a_j x_j - M}{S} \geqslant \tau_N\right) = \text{Prob}(\sum_j a_j x_j \geqslant M + \tau_N S) = \alpha$$

and

$$M + \tau_N S \geqslant b \quad \text{is equivalent to} \quad \text{Prob}(\sum_j a_j x_j \geqslant b) \geqslant \alpha.$$

Alternatively, if we take as probability densities

(ii) $$\frac{1}{\theta\pi\left[1 + \left(\dfrac{a_j - \mu_j}{\theta_j}\right)^2\right]}$$ [that of the Cauchy distribution, $(\theta_j > 0)$] then the characteristic function is

$$\exp(i\mu_j t - \theta_j |t|)$$

and hence that of $\sum_j a_j x_j$ is

$$\exp[it \sum_j \mu_j x_j - |t| \sum_j \theta_j x_j]$$

or

$$\text{(iii)} \quad \frac{1}{c_j\sqrt{2\pi}}\left(\frac{a_j - B_j}{c_j}\right)^{-3/2} \exp\left(-\frac{1}{2}\frac{c_j}{a_j - B_j}\right) \qquad \text{for} \quad a_j \geqslant B_j$$

and

$$0 \qquad \text{for} \quad a_j < B \quad (c_j > 0)$$

then the characteristic function is

$$\exp(iB_j t + \sqrt{c_j|t|}\,\{1 - \text{sgn}\,t\}).$$

This is a function studied by P. Lévy (1939), and also found by N. V. Smirnov.† The characteristic function of $\sum_j a_j x_j$ is then

$$\exp(it \sum_j B_j x_j + \sum_j \sqrt{c_j x_j|t|}\,\{1 - \text{sgn}\,t\}).$$

It follows that the deterministic equivalents in these two cases are, respectively,

$$\text{(ii)} \quad \sum_j \mu_j x_j + \tau_c \sum_j \theta_j x_j \geqslant b$$

where

$$\frac{1}{\pi}\int_{-\infty}^{\tau_c} \frac{dx}{1 + x^2} = 1 - \alpha$$

and

$$\text{(iii)} \quad \sum_j B_j x_j + \tau_L (\sum_j \sqrt{c_j x_j})^2 \geqslant b$$

† According to B. V. Gnedenko and A. N. Kolmogorov (1954).

where

$$\frac{1}{\sqrt{2\pi}} \int_{-\infty}^{\tau_L} x^{-3/2} \exp\left(\frac{-1}{2x}\right) dx = 1 - \alpha.$$

In all these cases we had distributions with two parameters, u and v, such that the convolution of two distribution functions, say $F\left(\frac{x-u_1}{v_1}\right)$ and $F\left(\frac{x-u_2}{v_2}\right)$, and hence that of any number of distribution functions, was again of the form $F\left(\frac{x-u}{v}\right)$. We knew how to obtain u from u_1 and u_2, and v from v_1 and v_2. Also, v, v_1, and v_2 were positive.

Such distribution functions are called stable. (Compare Lukacs, 1960.) The three which we mentioned are the only ones known whose distribution function can be expressed in terms of elementary functions.

It is clear that if b is also random and belongs to the same stable family as the a_j, then the same argument can be used for finding deterministic equivalents, by writing for $\sum_{j=1}^{m} a_j x_j \geqslant b$ the inequality $\sum_{j=1}^{m+1} a_j x_j \geqslant 0$ with $a_{m+1} = -b$, and $x_{m+1} = 1$.

The deterministic constraint which we found for the case of the normal distributions can be written

$$\sum_j \bar{a}_j x_j - b \geqslant -\tau_N (\sum_j \sigma_j^2 x_j^2)^{1/2}.$$

If $\alpha > \frac{1}{2}$, which will be the case in most applications, then τ_N is negative. We can then square both sides of the inequality and obtain

$$(\sum_j \bar{a}_j x_j)^2 - 2b \sum_j \bar{a}_j x_j - \tau_N^2 \sum_j \sigma_j^2 x_j^2 \geqslant b,$$

a quadratic constraint.

If the a_j have distributions which are not stable (and b is given), then we can try other methods to obtain deterministic equivalents. To begin with (cf. Charnes, Kirby, and Raike, 1969†) consider, for all j, all γ_j such that $\sum_{j=1}^{n} \gamma_j x_j \geqslant b$ for given x_j. Then the requirement

$$\text{Prob}(\sum_j a_j x_j \geqslant b) \geqslant \alpha$$

is equivalent to the requirement that the total probability of $a_j \geqslant \gamma_j$ for all such γ_j be at least α.

PROOF: Whenever $\sum_j a_j x_j \geqslant b$, we have also $\gamma_j (= a_j)$ $(j = 1, \ldots, n)$ for which

$$\sum_j \gamma_j x_j \geqslant b.$$

Conversely, if $\sum_j \gamma_j x_j \geqslant b$, then $\sum_j a_j x_j \geqslant \sum_j \gamma_j x_j \geqslant b$ follows. ■

However, this equivalence is not useful for computational purposes, and we give therefore a sufficient condition for the chance constraint to hold.

If, for given x, there exist γ_j $(j = 1, \ldots, n)$ such that

$$\sum_j \gamma_j x_j \geqslant b \tag{*}$$

and

$$\prod_j \text{Prob}(a_j \geqslant \gamma_j) \geqslant \alpha \tag{**}$$

† The expression "zero-zero" in the title of this paper indicates that it deals with zero-order decision rules for zero-sum games.

then

$$\text{Prob}(\sum_j a_j x_j \geqslant b) \geqslant \alpha.$$

PROOF: We have

$$\prod_j \text{Prob}(a_j \geqslant \gamma_j) \leqslant \prod_j \text{Prob}(a_j x_j \geqslant \gamma_j x_j)$$

(strict inequality arises when $x_j = 0$). Therefore, from (**)

$$\prod_j \text{Prob}(a_j x_j \geqslant \gamma_j x_j) = \text{Prob}(a_1 x_1 \geqslant \gamma_1 x_1, \ldots, a_n x_n \geqslant \gamma_n x_n) \geqslant \alpha$$

and hence

$$\text{Prob}(\sum_j a_j x_j \geqslant \sum_j \gamma_j x_j) \geqslant \alpha.$$

The result follows then from (*). ■

This condition is not necessary, though. For instance, let $b = 1$ and $x_1 = x_2 = 1$. Let a_1 as well as a_2 be uniformly and independently distributed in $(0, 1)$. Then

$$\text{Prob}(a_1 + a_2 \geqslant 1) = \tfrac{1}{2}.$$

But if $\gamma_1 + \gamma_2 = 1$, then no γ_1 and $\gamma_2 = 1 - \gamma_1$ exist such that

$$\text{Prob}(a_1 \geqslant \gamma_1) \cdot \text{Prob}(a_2 \geqslant 1 - \gamma_1) \geqslant \tfrac{1}{2}$$

because the maximum of the shaded area (see Figure 6) is reached when $\gamma_1 = \gamma_2 = \frac{1}{2}$ and is then only $\frac{1}{4}$, while for $\gamma_1 + \gamma_2 > 1$ the maximum area would be even smaller.

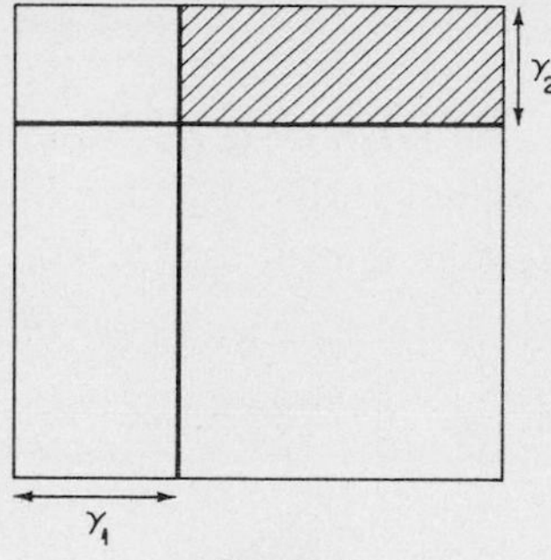

Figure 6

A necessary condition for $\text{Prob}(\sum_j a_j x_j \geqslant b) \geqslant \alpha > 0$ to hold, when $x_1 + \cdots + x_n = 1$, is the existence of $\delta_j \, (j = 1, \ldots, n)$ such that

$$\sum_j \delta_j x_j \geqslant b \qquad \text{and} \qquad 1 - \prod_j \text{Prob}(a_j < \delta_j) \geqslant \alpha.$$

PROOF: Let t be a parameter such that $\sum_j a_j(t) x_j \geqslant b$. Then for some j we have $a_j(t) \geqslant b$ (since $x_j \geqslant 0, \sum_j x_j = 1$). The choice $\delta_j = a_j(t)$ satisfies both inequalities: the first by definition of $a_j(t)$, and the second because the left-hand side is the probability that at least for one j we shall have $a_j \leqslant \delta_j$ holding with probability α.

If the a_j have a joint normal distribution with mean vector $\bar{a}$ and covariance matrix V, then $\sum_j a_j x_j$ is normally distributed with mean $M = \sum_j \bar{a}_j x_j$ and standard deviation $S = (x'Vx)^{1/2}$, so that if

$$\frac{1}{\sqrt{2\pi}} \int_{-\infty}^{\tau} \exp(-\tfrac{1}{2}x^2)\, dx = 1 - a$$

then

$$\text{Prob}(\sum_j a_j x_j \geqslant b) \geqslant \alpha \quad \text{is equivalent to} \quad M + \tau S \geqslant b.$$

Joint Probability

Consider now the case of minimizing $c'x$, subject to $x \geqslant 0$ and to the further constraint that the joint probability distribution of all inequalities $\sum_j a_{ij}x_j \geqslant b_i$ $(i = 1, \ldots, m)$ to hold be at least $\alpha > 0$.

If we still take the b_i to be independently distributed with several distribution functions $F_i(z_i) = \text{Prob}(b_i \leqslant z_i)$, then the condition can be written

$$F_1(b_1) \cdots F_m(b_m) \geqslant \alpha.$$

Such an example appears in Charnes and Stedry (1964), and its theory has been developed by Miller and Wagner (1965). [The more general case, when the b_i are not independent but restricted to two inequalities $\sum_j a_{ij}x_j \geqslant b_i$ $(i = 1, 2)$, has been dealt with by Balintfy (1970) and by A. Prékopa (1970) for the normal distribution.]

The feasible region is convex if $\prod_j F_i(b_i)$ is concave, and we shall now investigate when this is so. We show that if this product is concave, then every F_i must be concave.

PROOF: We assume, with $\lambda + \mu = 1$, $\lambda, \mu \geqslant 0$

$$F_i(\lambda b_{1i} + \mu b_{2i}) \prod_{j \neq i} F_j(b_j) \geqslant \lambda F_i(b_{1i}) \prod_{j \neq i} F_j(b_j) + \mu F_i(b_{2i}) \prod_{j \neq i} F_j(b_j)$$

where the b_j are fixed values. Because $\alpha > 0$, no $F_j(b_j)$ can be zero, and we can therefore divide by the indicated product, which proves the theorem for $F_i(b_i)$. ■

It follows, also, from an investigation of the second derivative of $\log F_i(b_i)$ that it is negative, when that of $F_i(b_i)$ is negative, so that $\log F_i(b_i)$ is also concave.

Miller and Wagner (1965) prove also other necessary conditions for concavity.

Many often used distribution functions, e.g. the normal, and the uniform distributions, are not concave, and neither is (therefore) any product of such functions. However, we can still use convex programming algorithms, if by taking logarithms we obtain a sum of concave functions. This is the case, for instance (cf. Miller and Wagner, 1965), if the probability densities are of the form

$$\left(\frac{n}{v-u}\right)\left(\frac{b-u}{v-u}\right)^{n-1} \qquad \text{for} \quad u \leqslant b \leqslant v$$

and

$$0 \qquad \text{otherwise.}$$

The uniform distribution is the special case with $n = 1$.

Example. Minimize

$$x_1 + x_2$$

subject to

$$\text{Prob}\left(\frac{x_1}{9} + x_2 \geqslant b_1, x_1 + \frac{x_2}{9} \geqslant b_2\right) = \tfrac{1}{8}$$

$$x_1, x_2 \geqslant 0$$

with b_i uniformly distributed in (0, 2), and b_2 uniform in (0, 4).

A deterministic equivalent can easily be constructed for this case, because of the uniform distribution of the b_i. It is

$$\frac{x_1}{9} + x_2 \geqslant u_1, \qquad x_1 + \frac{x_2}{9} \geqslant u_2$$

$$0 \leqslant u_1 \leqslant 2$$

$$0 \leqslant u_2 \leqslant 4$$

$$u_1 \cdot u_2 = 1.$$

Randomized Decisions

We show now by a simple example that if we allow randomized decisions, and minimize an expected value, then this value could be smaller than if we insisted on nonrandomized choices.

Let it be required to minimize Ex, subject to

$$x \geqslant 0 \qquad \text{and} \qquad \operatorname{Prob}(x \geqslant b) \geqslant \alpha$$

where x and b are scalars, and b is random in the interval $(0,1)$ with probability density $2b$. Of course, $\int_0^1 2b\,db = 1$. The deterministic equivalent to the chance constraint is $x^2 \geqslant \alpha$, so that the optimal $x_0 = \sqrt{\alpha} = Ex$.

However, if we choose x at random between 0 and 1, with some density $\beta(x)$, such that

$$\int_0^1 \int_0^x 2b\,db\,\beta(x)\,dx = \int_0^1 \beta(x)\,x^2\,dx = \alpha$$

then the chance constraint is satisfied, and $Ex = \int_0^1 \beta(x)\,x\,dx < \sqrt{\alpha}$, unless $\beta(x)$ is degenerate, because

$$\alpha = \int_0^1 \beta(x)\,x^2\,dx \geqslant \left[\int_0^1 \beta(x)\,x\,dx\right]^2 = (Ex)^2$$

since

$$\int_0^1 \beta(x)\,x^2\,dx - \left[\int_0^1 \beta(x)\,x\,dx\right]^2 = \int_0^1 \beta(x)\left[x - \int_0^1 \beta(x)\,x\,dx\right]^2 dx.$$

[This example is similar to one given by Dempster (1968) for a discretely distributed b.]

Such an example suggests a search for the best distribution $\beta(x)$, i.e. that which makes $Ec'x$ smaller than any other distribution.

We deal again with the example above, and assume, to begin with, that $\beta(x)$ is discontinuous, in that only the probabilities $p(x)$ of $x = i/n$ $(i = 0, 1, \ldots, n)$ are positive.

Then $Ex = \sum_{i=0}^{n} p(i/n)(i/n)$ is to be minimized, subject to

$$\sum_{i=0}^{n} p\left(\frac{i}{n}\right) = 1 \qquad \text{and} \qquad \sum_{i=0}^{n} p\left(\frac{i}{n}\right)\left(\frac{i}{n}\right)^2 = \alpha.$$

This is a linear programming problem with the $p(i/n)$'s as unknowns. We solve for $p(0)$ and $p(1)$ and have

$$p(0) = 1 - \alpha + \sum_{i=1}^{n-1} p\left(\frac{i}{n}\right)\left[\left(\frac{i}{n}\right)^2 - 1\right]$$

$$p(1) = \alpha - \sum_{i=1}^{n-1} p\left(\frac{i}{n}\right)\left(\frac{i}{n}\right)^2$$

and

$$Ex = \alpha + \sum_{i=1}^{n-1} p\left(\frac{i}{n}\right)\left[\left(\frac{i}{n}\right) - \left(\frac{i}{n}\right)^2\right].$$

The last term is positive (because i/n for $i \neq 0$ or n is a proper fraction) and it follows that the optimal solution is

$$p(0) = 1 - \alpha, \qquad p(1) = \alpha$$

$$\text{all other} \quad p\left(\frac{i}{n}\right) = 0 \qquad \text{and} \qquad Ex = \alpha.$$

This remains true if we increase n without bounds, so that even if continuous distributions are admitted, the best choice for x within 0 and 1 is to concentrate on $x = 0$ and $x = 1$, in proportions $1 - \alpha$ and α. Then $Ex = \alpha$, which is better than the previous $\sqrt{\alpha}$, if the probability α is a proper fraction.

The method we have used here is that of Greenberg (1968), who proves that Ex^2,

$$\text{subject to} \quad \text{Prob}(x \leqslant b) \geqslant \alpha \geqslant \tfrac{1}{2}$$

where

$$\text{Prob}(b = -1) = \text{Prob}(b = -2) = \tfrac{1}{2}$$

is minimized by the distribution

$$\text{Prob}(x = -1) = 2 - 2\alpha \qquad \text{and} \qquad \text{Prob}(x = -2) = 2\alpha - 1.$$

Greenberg (1968) deals also with the more general problem of minimizing Ex^2 subject to

$$p\,\text{Prob}(x = -1) + (1 - p)\,\text{Prob}(x = -2) \geqslant \alpha.$$

The example of Wessels (1967), see p. 76, concerns also such randomized decision rules.

P-Model

Charnes and Cooper (1963) call a program which optimizes the expectation of $c'x$ an *E*-model, while a *V*-model is one which minimizes the expected value of a quadratic expression, e.g. a variance. They add to this what they call a *P*-model, where it is required to maximize the probability α that $c'x$ does not exceed a given constant, k. The vector c is stochastic, and x is subject to constraints. Bereanu (1964a) called this a *program of minimal risk, at level k*, while Simon (1957) speaks of *satisficing* rather than optimizing an objective.

We shall again transform this problem into one where no probability enters, a *deterministic equivalent*.

If c can be written as $c_0 + tc_1$, where t is a scalar with a distribution function, then the probability of

$$c'x = c_0'x + tc_1'x \leqslant k$$

is that of

$$t \leqslant \frac{(k - c_0{}'x)}{c_1{}'x}$$

provided $c_1{}'x$ is always positive. In this case, therefore, we have to maximize the fraction of two linear functions, which is a known problem of *fractional linear programming*.

If the joint distribution of the components c_j of c is normal, with means m_j and covariance matrix V, then $c'x$ is normally distributed with mean $M = \sum m_j x_j$ and standard deviation $S = (x'Vx)^{1/2}$. The probability of $c'x$ not exceeding k is then

$$\frac{1}{S\sqrt{2\pi}} \int_{-\infty}^{k} \exp\left[-\frac{(y-M)^2}{2S^2} \right] dy = \frac{1}{\sqrt{2\pi}} \int_{-\infty}^{(k-M)/S} \exp(-\tfrac{1}{2}u^2)\, du$$

and this is maximized when $(k-M)/S$ is. We have thus once more a fractional objective function, though not the quotient of two linear expressions. If, by virtue of the constraints on x, we know that the numerator is positive, then we can square the quotient and have to maximize the ratio of two quadratic expressions.

Kataoka (1963), on the other hand, considers the minimization of k, subject to $\text{Prob}(c'x \leqslant k) = \alpha$.

In this case, assuming again that the components c_j of c have a joint normal distribution with means m_j and covariance matrix V, the chance constraint can be transformed into

$$k = \sum_j m_j x_j + \tau (x'Vx)^{1/2}$$

where τ is chosen so that

$$\frac{1}{\sqrt{2\pi}} \int_{-\infty}^{\tau} \exp(-\tfrac{1}{2}y^2)\, dy = \alpha.$$

We are thus led to minimizing the expression given above for k, and we can show that, if τ is positive (as it will be if $\alpha > \frac{1}{2}$), this expression is convex. We need, of course, only to prove this for $(x'Vx)^{1/2}$, or for $(x'Vx)^2$.

Now

$$(x'Vy)^2 \leqslant (x'Vx)(y'Vy)$$

by Schwartz's inequality, since V is positive semidefinite. This can be written (with $\lambda+\mu = 1$, $\lambda, \mu \geqslant 0$)

$$[\mu y' V \lambda x + \lambda x' V \mu y]^2 \leqslant 2\lambda\mu x'Vx \cdot y'Vy$$

or

$$\{[\lambda x + \mu y]' V [\lambda x + \mu y]\}^2 \leqslant \{\lambda x'Vx + \mu y'Vy\}^2. \quad \blacksquare$$

An approach, again different from those so far discussed, requires to find an x which maximizes the probability that the minimum of $c'x+d'y$

subject to

$$Ax + By = b$$

$$x, y \geqslant 0$$

does not exceed k (a given value). This is a problem which belongs into the field of two-stage programs.

To illustrate this case, we use again the example:

Minimize

$$x + 2y$$

subject to

$$0 \leqslant x \leqslant 100, \qquad x + y \geqslant b, \qquad x, y \geqslant 0$$

where b is uniformly distributed, this time, between 70 and 120, and demand that the probability of that minimum not exceeding 110 be as large as possible. [Discussed by A. Madansky (1959).]

We have seen that the minimum of $x+2y$, subject to the constraints, equals

$$x \quad \text{if} \quad b \leqslant x, \qquad \text{and} \qquad x+2(b-x) = 2b-x \quad \text{if} \quad b \geqslant x.$$

Hence

$$C(b,x) \leqslant 110 \qquad \text{when} \quad b \leqslant x \qquad \text{and} \qquad x \leqslant 110$$

and also

$$\text{when} \quad b \geqslant x \qquad \text{and} \qquad 2b-x \leqslant 110$$

$$\text{i.e.} \quad b \leqslant 55+\tfrac{1}{2}x.$$

The solution of our program is easily seen in the following diagram (Figure 7). The largest vertical in the triangle ABC is that for $x = 100$ leading from $b = 70$ to $b = 105$. Hence the maximum probability is $35/50 = 0.7$.

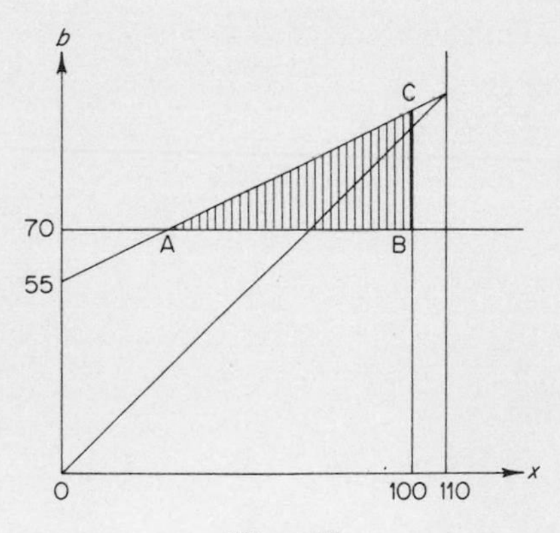

Figure 7

We consider also a nonlinear example of this type, from Vajda (1970). Find an $x \geqslant 0$ which maximizes the probability that the

$$\text{minimum of} \quad x^2 + 2y^2$$

does not exceed

$$k = 12100$$

subject to

$$0 \leqslant x \leqslant 100, \qquad x+y \geqslant b, \qquad y \geqslant 0$$

where b is uniformly distributed in the interval (70, 120).
The minimum of

$$x^2 + 2y^2$$

is

$$x^2 \qquad \text{when} \quad b \leqslant x$$

and

$$x^2 + 2(b-x)^2 \qquad \text{when} \quad b \geqslant x.$$

therefore the minimum will not exceed k, when

$$\text{either} \quad b \leqslant x \quad \text{and} \quad x^2 \leqslant k, \quad \text{i.e.} \quad x \leqslant 110$$

$$\text{or} \quad b \geqslant x \quad \text{and} \quad x^2 + 2(b-x)^2 \leqslant k.$$

Figure 8 is drawn with x as abscissa and b as ordinate. Because b is uniformly distributed, we want to find that x with the largest vertical whose points satisfy

$$70 \leqslant b \leqslant 120, \qquad 0 \leqslant x \leqslant 100$$

and either $b \leqslant x$ and $x \leqslant 110$, or $b \geqslant x$ and $x^2+2(b-x)^2 \leqslant 110^2$.

The ellipse $x^2+2(b-x)^2 = 110^2$ intersects the line $b = 120$ in points where $x = 80 \pm 50/\sqrt{3}$, i.e. where $x = 51.1$ or 108.9. The latter value exceeds 100, and we see that the probability of $\min(x^2+2y^2)$ equals unity (the largest possible value) for all x between 51.1 and 100.

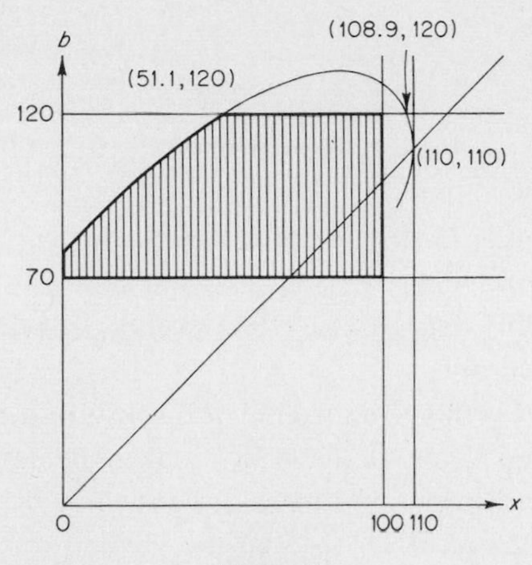

Figure 8

Other approaches have been suggested in the literature, some without any convenient algorithmic methods. Thus in Charnes, Cooper, and Thompson (1965b), it is required to find that vector x which makes the probability p that all constraints $\sum_j a_{ij}x_j \geqslant b_i$ be satisfied with at least p, as high as possible. The vector x might be subject to further constraints.

For instance, let x_1 and x_2 be nonnegative and let the chance constraints be

$$\text{Prob}(-2x_1 + x_2 \geqslant b_1) \geqslant p \qquad \text{and} \qquad \text{Prob}(x_1 - x_2 \geqslant b_2) \geqslant p$$

where b_1 is normally distributed with mean 1 and standard deviation 1, and b_2 with mean $-\frac{3}{2}$ and standard deviation 1. The chance constraints are equivalent to

$$-2x_1 + x_2 \geqslant 1 + t, \qquad x_1 - x_2 \geqslant -\tfrac{3}{2} + t$$

where t is to be maximized. The linear program is solved by

$$x_1 = 0, \qquad x_2 = \tfrac{5}{4}, \qquad t = \tfrac{1}{4}$$

which means that the (maximum obtainable) probability is

$$p = \frac{1}{\sqrt{2\pi}} \int_{-\infty}^{1/4} \exp(-\tfrac{1}{2}x^2)\, dx \simeq 0.5987.$$

Evers (1967) adds to the objective function (to be minimized) some multiple of the probability that at least one constraint is not satisfied, and this probability depends itself on the choice of x. In his study b as well as A are stochastic.

Sengupta and Portillo-Campbell (1970) speak of a fractile approach, when it is required to find that x which makes the limit, below which a fraction α of the distribution of the objective function lies, as large as possible. For instance, if $\alpha = \frac{1}{2}$, then the median (or, for a symmetric distribution, the mean) is to be maximized by the choice of x.

We might also think of the problem of finding that x which maximizes the probability of x being feasible. Examples in Chapter I show that this objective function is not necessarily concave.

For approximate solutions to chance-constrained programs we refer to Sengupta (1969).

Nonzero Order Rules

The decision rules—rules for finding the optimal x_j—which we have so far considered are *zero-order rules*; they determine x_j before the actual values of the random elements become known.

Another method consists in waiting for these values to become known, but deciding in advance how this knowledge is going to be used. This is the case, for instance, in the active approach of Tintner, mentioned in Chapter II, though there the constraints were to hold with probability 1.

When we have chance-constraints, we can try to replace them by other equivalent constraints, required to hold with probability 1, but still referring to chance elements (e.g. b_i). These constraints are called *certainty equivalents* (Charnes and Cooper, 1963, p. 22), because they do not imply any randomized decisions; for zero order rules they are the same as deterministic equivalents.

For example, take again the problem of minimizing $Ec'x$, subject to

$$\text{Prob}\left(\sum_j a_j x_j \geqslant b_i\right) \geqslant \alpha_i \qquad (i = 1, \ldots, m)$$

where the b_i are stochastic, but not the c_j and a_{ij}. The objective function is an expected value, because the x_j are themselves random, being dependent on the random b_i.

We might decide to use a linear decision rule $x = Db$, i.e.

$$x_j = \sum_{k=1}^{m} d_{jk} b_k$$

where the d_{jk} have to be determined optimally. (Compare Charnes and Cooper, 1963.)

Historically, the first application of this idea occurred in Charnes, Cooper, and Symonds (1958) in scheduling heating oil to an uncertain demand, where the matrix (a_{ij}) was lower triangular, and so was (d_{jk}).

The objective function can now be written

$$E\, c'Db = c'D\, Eb = c'D \cdot \bar{b}$$

and the constraints become

$$\text{Prob}\left(\sum_j a_{ij} \sum_k d_{jk} b_k - b_i \geqslant 0\right) \geqslant \alpha_i.$$

The distribution of each b_i is known. Assume, in particular, that

$$\sum_j a_{ij} \sum_k d_{jk} b_k - b_i$$

is a normally distributed random variable k_i, with mean $\bar{k}_i$ and standard deviation s_i. We want then to find τ_i such that

$$\operatorname{Prob}(k_i \geqslant \bar{k}_i + \tau_i s_i) = \alpha_i$$

so that the equivalents of the chance constraints are

$$\bar{k}_i + \tau_i s_i \geqslant 0.$$

Explicitly, this means, writing A_i for $(a_{i1}, \ldots, a_{in})$

$$A_i' D\bar{b} - \bar{b}_i + \tau_i [E(A_i' Db - b_i)^2 - (A_i' D\bar{b} - \bar{b}_i)^2]^{1/2} \geqslant 0$$

or, after introducing a nonnegative variable v_i

$$A_i' D\bar{b} - \bar{b}_i \geqslant v_i \geqslant -\tau_i [E(A_i' Db - b_i)^2 - (A_i' D\bar{b} - \bar{b}_i)^2]^{1/2}.$$

If $\alpha_i \geqslant \frac{1}{2}$, then the last term of this relation is nonnegative. We can then square the two sides of the second inequality, thus

$$v_i^2 + \tau_i^2 (A_i' D\bar{b} - \bar{b}_i)^2 - \tau_i^2 E(A_i' Db - b_i)^2 \geqslant 0.$$

This quadratic constraint, together with the linear

$$A' D\bar{b} - \bar{b}_i \geqslant v_i$$

replaces $\operatorname{Prob}(A_i' Db - b_i \geqslant 0) \geqslant \alpha_i$.

The variable v_i was introduced in order to show that the constraints form a convex region. Indeed, a region defined by a quadratic constraint of the form

$$v^2 + s^2 w^2 - t^2 u^2 \geqslant 0$$

and by $v \geqslant 0$, where the coordinates are v, w, and u, forms a nappe of an elliptic hyperboloid and its interior, which is a convex region.

The V-model and the P-model can be treated in a similar fashion.

Conditional Quantiles

We have mentioned that the first example of an application of a linear decision rule concerned a lower triangular matrix. As an example of such a case consider the following problem (cf. Charnes and Kirby, 1966):

Minimize

$$Ec'x$$

subject to

$$\begin{aligned} \mathrm{Prob}(a_{11}x_1 \geqslant b_1) &\geqslant \alpha_1 \\ \mathrm{Prob}(a_{21}x_1 + a_{22}x_2 \geqslant b_2) &\geqslant \alpha_2 \\ &\vdots \\ \mathrm{Prob}(a_{n1}x_1 + \cdots + a_{nn}x_n \geqslant b_n) &\geqslant \alpha_n \end{aligned}$$

where the b_i have a known joint probability distribution.

Such problems will, for instance, arise if b_i is the (still unknown) requirement for some commodity in the ith period, and the requirements must be satisfied by activity levels in all periods up to and including the ith.

In such a situation we need not decide on the precise values of all the x_j at the beginning of the first period, but need only determine x_1, and then x_2 at the start of the second period, when x_1 as well as b_1 are known, and so on. In general, we choose x_j when $b_1, \ldots, b_{j-1}$ have been observed, and $x_1, \ldots, x_{j-1}$ have already been determined.

We are now concerned with the conditional probabilities of

$$a_{j1}x_1 + \cdots + a_{jj}x_j \geqslant b_j, \qquad \text{given} \quad b_1, \ldots, b_{j-1}$$

(which values had an influence on the values $x_1, \ldots, x_{j-1}$), and we assume (though this might be difficult in any given case) that we can find $\bar{\alpha}_j$ such that the jth chance constraint is equivalent to

$$\text{Cond Prob}(a_{j1} x_1 + \cdots + a_{jj} x_j \geqslant b_j) \geqslant \bar{\alpha}_j .$$

If the joint distribution of all b_j is known, then so is the conditional distribution function of every b_j, conditional on $b_1, \ldots, b_{j-1}$, and we can find, as has been shown, equivalents of the chance constraints.

A randomized solution (as opposed to a certainty equivalent) to a problem similar to the last one was considered in Charnes and Cooper (1959).

Let b_i have discrete distributions and take the following problem:

Minimize

$$Ec'x$$

subject to

$$\text{Prob}(x_1 \geqslant b_1) \geqslant \alpha_1$$

$$\text{Prob}(x_1 + x_2 - b_1 \geqslant b_2) \geqslant \alpha_2$$

$$\vdots$$

$$\text{Prob}(x_1 + x_2 + \cdots + x_n - b_1 - \cdots - b_{n-1} \geqslant b_n) \geqslant \alpha_n .$$

Let $x_1 + \cdots + x_j - b_1 - \cdots - b_{j-1} = u_j$. ($u_1 = x_1$) is statistically independent of b_j, and the distribution of $u_j - b_j$ is the convolution of the distributions of u_j and of $-b_j$.

To make sure that no x_j turns out to be negative, in other words to ensure that $u_j \geqslant u_{j-1} - y_{j-1}$, or

$$\min u_j \geqslant \max(u_{j-1} - y_{j-1})$$

we might proceed as follows:

Let the values which u_j, or $u_{j-1} - y_{j-1}$ can take be, in ascending order

$$t_1, t_2, \ldots, t_s, \ldots.$$

Denote the probabilities of u_j taking any of these values by

$$p_1, p_2, \ldots, p_s, \ldots p_N$$

and those of $u_{j-1} - y_{j-1}$ taking any of these values by

$$q_1, q_2, \ldots, q_s, \ldots q_N.$$

(Of course, some of these values will be zero, but $\sum p_i = \sum q_i = 1$.)

We want to formulate the condition that either $\sum_1^s p_i = 0$ or $\sum_{s+1}^N q_i = 0$, for all s. To do this, we introduce a variable h_s with value either 0 or 1, and write

$$\sum_1^s p_i \leqslant h_s, \qquad \sum_1^s q_i \geqslant h_s.$$

This gives the correct answer, for when $h_s = 0$, then $p_1 = \cdots p_s = 0$ and $q_1, \ldots, q_s$ can have any value in $(0, 1)$. On the other hand, when $h_s = 1$, then $\sum_{s+1}^n q_i \leqslant 0$, i.e. $q_{s+1} = \cdots q_N = 0$, and $p_1, \ldots, p_s$ can have any value in $(0, 1)$.

This procedure must be repeated for all s.

Appendix I

Linear Programming and Duality

Linear programming is concerned with the following type of problems:

Minimize

$$x_0 = \sum_{j=1}^{n} c_j x_j$$

subject to

$$\sum_{j=1}^{n} a_{ij} x_j - x_{n+i} = b_i \qquad (i = 1, \ldots, m)$$

$$x_j, x_{n+i} \geqslant 0.$$

A set of m variables is called a basic set (and the set of the remaining variables a nonbasic set) if their matrix of coefficients is not singular. Without restricting generality we may assume that

$$x_1, \ldots, x_{m-s}, x_{n+1}, \ldots, x_{n+s}$$

form a basic set, so that the matrix

$$\begin{bmatrix} a_{11} & \cdots & a_{1\,m-s} & -1 & 0 & \cdots & 0 \\ & & \vdots & & & & \\ a_{s1} & \cdots & a_{s\,m-s} & 0 & 0 & \cdots & -1 \\ a_{s+1\,1} & \cdots & a_{s+1\,m-s} & 0 & 0 & \cdots & 0 \\ & & \vdots & & & & \\ a_{m1} & \cdots & a_{m\,m-s} & 0 & 0 & \cdots & 0 \end{bmatrix}$$

is not singular, i.e. the determinant

$$\begin{vmatrix} a_{s+1\,1} & a_{s+1\,m-s} \\ & \vdots \\ a_{m\,1} & a_{m\,m-s} \end{vmatrix}$$

is not zero.

The basic variables can then be expressed, uniquely, in terms of the nonbasic ones. We have

$$\sum_{j=1}^{m-s} a_{ij} x_j - x_{n+i} = b_i - \sum_{j=m-s+1}^{n} a_{ij} x_j \qquad (i = 1, \ldots, s)$$

$$\sum_{j=1}^{m-s} a_{ij} x_j = b_i - \sum_{j=m-s+1}^{n} a_{ij} x_j + x_{n+i} \qquad (i = s+1, \ldots, m)$$

which we write, in matrix notation:

$$\begin{bmatrix} A_{11} & -I_s \\ A_{21} & 0 \end{bmatrix} \begin{bmatrix} x^{(1)} \\ s^{(1)} \end{bmatrix} = \begin{bmatrix} b^{(1)} \\ b^{(2)} \end{bmatrix} - \begin{bmatrix} A_{12} & 0 \\ A_{22} & -I_{m-s} \end{bmatrix} \begin{bmatrix} x^{(2)} \\ s^{(2)} \end{bmatrix}$$

where

$$x^{(1)} = (x_1, \ldots, x_{m-s})' \qquad x^{(2)} = (x_{m-s+1}, \ldots, x_n)'$$

$$s^{(1)} = (x_{n+1}, \ldots, x_{n+s})' \qquad s^{(2)} = (x_{n+s+1}, \ldots, x_{n+m})'$$

$$b^{(1)} = (b_1, \ldots, b_s)' \qquad b^{(2)} = (b_{s+1}, \ldots, b_m)'.$$

I_k is the identity matrix of order k, and the other matrices have the appropriate dimensions.

Premultiplying both sides by

$$\begin{bmatrix} 0 & A_{21}^{-1} \\ -I_s & A_{11} A_{21}^{-1} \end{bmatrix}$$

we obtain

$$\begin{bmatrix} x^{(1)} \\ s^{(1)} \end{bmatrix} = \begin{bmatrix} A_{21}^{-1} b^{(2)} \\ -b^{(1)} + A_{11} A_{21}^{-1} b^{(2)} \end{bmatrix}$$

$$- \begin{bmatrix} A_{21}^{-1} A_{22} & -A_{21} \\ -A_{12} + A_{11} A_{21}^{-1} A_{22} & -A_{11} A_{21}^{-1} \end{bmatrix} \begin{bmatrix} x^{(2)} \\ s^{(2)} \end{bmatrix}$$

so that

$$
\begin{aligned}
x_0 &= c^{(1)\prime}x^{(1)} + c^{(2)\prime}x^{(2)} \\
&= c^{(1)\prime}A_{21}^{-1}b^{(2)} - (c^{(1)\prime}A_{21}^{-1}A_{22} - c^{(2)\prime})x^{(2)} + c^{(1)\prime}A_{21}^{-1}s^{(2)}.
\end{aligned}
$$

Here

$$c^{(1)\prime} = (c_1, \ldots, c_{m-s}) \qquad \text{and} \qquad c^{(2)\prime} = (c_{m-s+1}, \ldots, c_n).$$

If we put the nonbasic variables equal to zero, i.e. $x^{(2)} = 0$ and $s^{(2)} = 0$, then

$$x^{(1)} = A_{21}^{-1}b^{(2)} \qquad \text{and} \qquad s^{(1)} = -b^{(1)} + A_{11}A_{21}^{-1}b^{(2)}.$$

If these values are nonnegative,

$$A_{21}^{-1}b^{(2)} \geqslant 0 \tag{PF1}$$

$$A_{11}A_{21}^{-1}b^{(2)} - b^{(1)} \geqslant 0 \tag{PF2}$$

then the basic variables are *feasible*.

The value of the objective function is

$$x_o = c^{(1)\prime}A_{21}^{-1}b^{(2)}.$$

If

$$-c^{(1)\prime}A_{21}^{-1}A_{22} + c^{(2)\prime} \geqslant 0 \tag{PO1}$$

and

$$c^{(1)\prime}A_{21}^{-1} \geqslant 0 \tag{PO2}$$

then x_o can not be made smaller by an increase of the nonbasic variables from their present zero value, so that the basic variables are then optimal.

If this is not so, then the Simplex Method provides rules for making one nonbasic variable basic, and vice versa, and it can be proved that the optimum, provided it is finite, will be reached after a finite number of such iterations.

Next, consider the problem which is "dual" to that just considered (which was the "primal" problem).

Maximize

$$y_0 = \sum_{i=1}^{m} b_i y_{n+}$$

subject to

$$\sum_{i=1}^{m} a_{ij} y_{n+i} + y_j = c_j \qquad (j = 1, \ldots, n)$$

$$y_{n+i}, y_j \geqslant 0.$$

Either of these two problems can be considered as being the primal, and the other is then the dual.

If the variables

$$x_1, \ldots, x_{m-s}, x_{n+1}, \ldots, x_{n+s}$$

are basic in the primal problem, then the variables

$$y_{n+s+1}, \ldots, y_{n+m}, y_{m-s+1}, \ldots, y_m$$

are basic in the dual problem, because their matrix of coefficients

$$\begin{bmatrix} a_{s+1\ 1} & \cdots & a_{m1} & 0 & 0 & \cdots & 0 \\ & & \vdots & & & & \\ a_{s+1\ m-s} & \cdots & a_{m\ m-s} & 0 & 0 & \cdots & 0 \\ a_{s+1\ m-s+1} & \cdots & a_{m\ m-s+1} & 1 & 0 & \cdots & 0 \\ & & \vdots & & & & \\ a_{s+1\ n} & \cdots & a_{mn} & 0 & 0 & \cdots & 1 \end{bmatrix}$$

is not singular, since its determinant has the same value as that of the basic variables in the primal problem.

We have now

$$\sum_{i=s+1}^{m} a_{ij} y_{n+i} = c_j - \sum_{i=1}^{s} a_{ij} y_{n+i} - y_j \qquad (j = 1, \ldots, m-s)$$

$$\sum_{i=s+1}^{m} a_{ij} y_{n+i} + y_j = c_j - \sum_{i=1}^{s} a_{ij} y_{n+i} \qquad (j = m-s+1, \ldots, n)$$

or, in matrix notation

$$\begin{bmatrix} A'_{21} & 0 \\ A'_{22} & I_{n-m+s} \end{bmatrix} \begin{bmatrix} y^{(1)} \\ t^{(1)} \end{bmatrix} = \begin{bmatrix} c^{(1)} \\ c^{(2)} \end{bmatrix} - \begin{bmatrix} A'_{11} & I_{m-s} \\ A'_{12} & 0 \end{bmatrix} \begin{bmatrix} y^{(2)} \\ t^{(2)} \end{bmatrix}$$

where

$$y^{(1)} = (y_{n+s+1}, \ldots, y_{n+m})'$$

$$t^{(1)} = (y_{m-s+1}, \ldots, y_m)'$$

and the meaning of $y^{(2)}$ and $t^{(2)}$ is obvious. We note, in particular, that $A_{11}, A_{12}, A_{21}, A_{22}$ appear again in their appropriate places, but transposed.

Premultiplying both sides by

$$\begin{bmatrix} (A'_{21})^{-1} & 0 \\ -A'_{22}(A'_{21})^{-1} & I_{n-m+s} \end{bmatrix}$$

we obtain

$$\begin{bmatrix} y^{(1)} \\ t^{(1)} \end{bmatrix} = \begin{bmatrix} (A')^{-1} c^{(1)} \\ -A'_{22}(A'_{21})^{-1} c^{(1)} + c^{(2)} \end{bmatrix}$$

$$- \begin{bmatrix} (A'_{21})^{-1} A'_{11} & (A'_{21})^{-1} \\ -A'_{22}(A'_{21})^{-1} A'_{11} + A'_{12} & -A'_{22}(A'_{21})^{-1} \end{bmatrix} \begin{bmatrix} y^{(2)} \\ t^{(2)} \end{bmatrix}$$

so that

$$\begin{aligned} y_o &= b^{(1)\prime} y^{(2)} + b^{(2)\prime} y^{(1)} \\ &= b^{(2)\prime}(A'_{21})^{-1} c^{(1)} + (b^{(1)\prime} - b^{(2)\prime}(A'_{21})^{-1} A'_{11}) y^{(2)} - b^{(2)\prime}(A'_{21})^{-1} t^{(2)}. \end{aligned}$$

If we now put the nonbasic variables equal to zero, i.e. $y^{(2)} = 0$, and $t^{(2)} = 0$, then we find that the basic variables are feasible if

$$(A'_{21})^{-1} c^{(1)} \geqslant 0 \qquad \text{(DF1)}$$

$$-A'_{22}(A'_{21})^{-1} c^{(1)} + c^{(2)} \geqslant 0. \qquad \text{(DF2)}$$

The value of y_o is $b^{(2)\prime}(A'_{21})^{-1} c^{(1)}$ and it is maximal if

$$-b^{(1)\prime} + b^{(2)\prime}(A'_{21})^{-1} A'_{11} \geqslant 0 \qquad \text{(DO1)}$$

$$b^{(2)\prime}(A)'_{21}{}^{-1} \geqslant 0. \qquad \text{(DO2)}$$

The left-hand sides of (PF1) and (PF2) are, respectively, the transposes of those of (DO2) and (DO1), while the left hand sides of (DF1) and (DF2) are the transposes of those of (PO2) and (PO1). Hence, if the basic variables

$$x_1, \ldots, x_{m-s}, x_{n+1}, \ldots, x_{n+s}$$

are feasible and optimal for the primal problem, then so are

$$y_{m-s+1}, \ldots, y_m, y_{n+s+1}, \ldots, y_{n+m}$$

for the dual problem.

Moreover, we see that the two optimal values of the objective functions are equal

$$c^{(1)\prime} A_{21}^{-1} b^{(2)} = b^{(2)\prime} (A_{21}')^{-1} c^{(1)}.$$

A comparison of the subscripts of the basic optimal variables in the two problems shows that if y_t is basic and optimal in its problem, then x_t is nonbasic at the optimum in its own. If we indicate the optimal values by a superscript $^{\circ}$, then we can write this:

$$x_t^{\circ} y_t^{\circ} = 0 \qquad (t = 1, \ldots, n, n+1, \ldots, n+m).$$

This relationship, called "complementary slackness," follows immediately from the equality of the two optimal values. To see this, consider the expression

$$\sum_{i=1}^{m} \left(\sum_{j=1}^{n} a_{ij} x_j^{\circ} - b_i \right) y_{n+i}^{\circ} + \sum_{j=1}^{n} \left(c_j - \sum_{j=1}^{m} a_{ij} y_{n+i}^{\circ} \right) x_j^{\circ}$$

$$= \sum_{i=1}^{m} x_{n+i}^{\circ} y_{n+i}^{\circ} + \sum_{j=1}^{n} x_j^{\circ} y_j^{\circ}.$$

All terms in this expression are nonnegative, in view of the constraints of the two problems. However, if we write it

$$\sum_{i=1}^{m} \sum_{j=1}^{n} a_{ij} x_j^{\circ} y_{n+i}^{\circ} - \sum_{j=1}^{n} \sum_{i=1}^{m} a_{ij} y_{n+i}^{\circ} x_j^{\circ} + \sum_{j=1}^{n} c_j x_j^{\circ} - \sum_{i=1}^{m} b_i y_{n+i}^{\circ}$$

then the first two terms cancel, and so do the last two, because they are the optimal values of the two objective functions. Hence every term $x_t^{\circ} y_t^{\circ}$ must be zero.

These results were obtained assuming that the optimum value of one of the objective functions exists (i.e. is finite) and hence that of its dual exists as well. We proceed to see what happens if this assumption is not justified.

From the constraints

$$\sum_{j=1}^{n} a_{ij} x_j \geqslant b_i \qquad \sum_{i=1}^{m} a_{ij} y_{n+i} \leqslant c_j$$

$$x_j \geqslant 0 \qquad y_{n+i} \geqslant 0$$

it follows that

$$\sum_{j=1}^{n} x_j c_j \geqslant \sum_{i=1}^{m} \sum_{j=1}^{n} a_{ij} x_j y_{n+i} \geqslant \sum_{i=1}^{m} y_{n+i} b_i$$

so that for all x_j and y_{n+i} satisfying the constraints, we have

$$\min \sum_{j=1}^{n} x_j c_j \geqslant \max \sum_{i=1}^{m} y_{n+i} b_i .$$

Hence, if one of these two optima is infinite ($-\infty$ or $+\infty$, respectively), then the other can not be infinite as well, and it follows that the constraints of the other problem must then be contradictory. (It is, in fact, possible that each of the two systems of constraints is contradictory.)

Appendix II

Applications of Stochastic (Probabilistic) Programming in Various Fields

(References)

AGRICULTURE

Tintner, G., 1955
Tintner, G., 1960
Sengupta, J. K., G. Tintner, and C. Millham, 1963
Sengupta, J. K., G. Tintner, and B. Morrison, 1963
Van Moeseke, P., 1965

AIRCRAFT SCHEDULING

Dantzig, G. B. and A. R. Ferguson, 1956
Midler, J. L. and R. D. Wollmer, 1969

CHEMICAL INDUSTRY

Charnes, A., W. W. Cooper, and G. H. Symonds, 1958

CONTROL THEORY

Van Slyke, R. and R. Wets, 1966

FINANCE

Agnew, N. H., R. A. Agnew, J. Rasmussen, and K. R. Smith, 1969
Byrne, R., A. Charnes, W. W. Cooper, and K. Kortanek, 1967
Byrne, R., A. Charnes, W. W. Cooper, and K. Kortanek, 1968
Charnes, A., W. W. Cooper, and G. L. Thompson, 1965b
Charnes, A. and M. J. L. Kirby, 1965
Charnes, A. and S. Thore, 1966
Naslund, B. and A. Whinston, 1962
Van Moeseke, P., 1965

MANAGEMENT

Charnes, A., W. W. Cooper, and G. J. Thompson, 1964
Charnes, A. and A. C. Stedry, 1964
Charnes, A. and A. C. Stedry, 1966

MARKETING

Charnes, A., W. W. Cooper, J. K. Devoe, and D. B. Learner, 1966
Charnes, A., W. W. Cooper, J. K. Devoe, and D. B. Learner, 1968

NETWORKS

Charnes, A., M. J. L. Kirby, and W. M. Raike, 1966
Midler, J. L. 1970

NUTRITION

Van de Panne, C. and W. Popp, 1963

TRANSPORT

Charnes, A. and W. W. Cooper, 1959
Charnes, A. and W. W. Cooper, 1960
Williams, A. C., 1963
Szwarc, W., 1964
Charnes, A., J. Drèze, and M. Miller, 1966

WAREHOUSING

Charnes, A., J. Drèze, and M. Miller, 1966

REFERENCES

Agnew, N. H., R. A. Agnew, J. Rasmussen, and K. R. Smith (1969). An application of chance constrained programming to portfolio selection in a casualty insurance firm. *Man. Sci.* **15B**, 512–520.

Babbar, M. M. (1955). Distributions of solutions of a set of linear equations (with an application to linear programming). *J. Amer. Statist. Assoc.* **50**, 854–869.

Balintfy, J. L. (1970). Nonlinear programming for models with joint chance constraints. In "Integer and Nonlinear Programming" (J. Abadie, ed.). North-Holland, Publ., Amsterdam.

Beale, E. M. L. (1955). On minimizing a convex function subject to linear inequalities. *J. Roy. Statist. Soc.* **B17**, 173–184.

Beale, E. M. L. (1961). The use of quadratic programming in stochastic linear programming. *Rand Report*, P 2404.

Bereanu, B. (1963a). On stochastic programming I. Distribution problems: A single random variable. *Rev. Math. Pures Appl.* (Acad. de la Rep. Pop. Roumaine). **8**, 683–697.

Bereanu, B. (1963b, c). Problema stocastică a transportului: I. Cazul costurilor aleatoare; II Cazul consumurilor aleatoare. *Com. Acad. Rep. Pop. Romîne* **13**, 325–331, 333–337.

Bereanu, B. (1964a). Programme de risque minimal en programmation linéaire stochastique. *C. R. Acad. Sci. Paris* **259**, 981–983.

Bereanu, B. (1964b). Régions de décision et répartition de l'optimum dans la programmation linéaire. *C. R. Acad. Sci. Paris* **259**, 1383–1386.

Bereanu, B. (1966). On stochastic linear programming. The Laplace Transform of the optimum and applications. *J. Math. Anal. Appl.* **15**, 280–294.

Bereanu, B. (1967). On stochastic linear programming. Distribution problems, stochastic technology matrix. *Z. Wahrsch. verw. Geb.* **8**, 148–152.

Bui Trong Lieu. (1964). On a problem of convexity and its applications to non-linear stochastic programming. *J. Math. Anal. Appl.* **8**, 177–187.

Byrne, R., A. Charnes, W. W. Cooper, and K. Kortanek (1967). A chance constrained approach to capital budgeting. *J. Fin. Quantit. Anal.* **II**, 339–364.

Byrne, R., A. Charnes, W. W. Cooper, and K. Kortanek (1968). Some new approaches to risk. *Accounting Rev.* **43**, 18–37.

Charnes, A. and W. W. Cooper (1959). Chance constrained programming. *Man. Sci.* **6**, 73–79.

Charnes, A. and W. W. Cooper (1960). Chance constrained programs with normal deviates and linear decision rules. *Nav. Res. Logist. Quart.* **7**, 533–544.

Charnes, A. and W. W. Cooper (1962). Chance constraints and normal deviates. *J. Amer. Statist. Assoc.* **57**, 134–148.

Charnes, A. and W. W. Cooper (1963). Deterministic equivalents for optimizing and satisficing under chance constraints. *Operations Res.* **11**, 18–39.

Charnes, A., W. W. Cooper, J. K. Devoe, and D. B. Learner (1966). DEMON. Decision mapping via optimum go-no networks. A model for marketing new products. *Man. Sci.* **12**, 865–887.

Charnes, A., W. W. Cooper, J. K. Devoe, and D. B. Learner (1968). DEMON Mark II. External equations solution and approximation. *Man. Sci.* **14**, 682–691.

Charnes, A., W. W. Cooper, and G. H. Symonds (1958). Cost horizons and certainty equivalents: An approach to stochastic programming of heating oil. *Man. Sci.* **4**, 235–263.

Charnes, A., W. W. Cooper, and G. J. Thompson (1964). Critical path analysis via chance constrained and stochastic programming. *Operations Res.* **12**, 460–470.

Charnes, A., W. W. Cooper, and G. L. Thompson (1965a). Constrained generalized medians and hypermedians as deterministic equivalents for two-stage linear programs under uncertainty. *Man. Sci.* **12**, 83–112.

Charnes, A., W. W. Cooper, and G. L. Thompson (1965b). Chance-constrained programming and related approaches to cost effectiveness. *System Res. Memorandum* 123. Northwestern University, Evanston, Illinois.

Charnes, A., J. Drèze, and M. Miller (1966). Decision and horizon rules for stochastic planning problems: A linear example. *Econometrica* **34**, 307–330.

Charnes, A. and M. J. L. Kirby (1965). Application of chance-constrained programming to solution of the so-called "Savings and

Loan Association" type of problem. *Res. Anal. Corp.* RAC-P-12.

Charnes, A. and M. Kirby (1966). Optimal decision rules for the *E*-model of chance-constrained programming. *Cah. Centre Etudes Rech. Oper.* **8**, 5–44.

Charnes, A., M. J. L. Kirby, and W. M. Raike (1966). Chance-constrained generalized network. *Operations Res.* **14**, 1113–1120.

Charnes, A. and M. J. L. Kirby (1967). Some special *P*-models in chance-constrained programming. *Man. Sci.* **14**, 183–195.

Charnes, A., M. J. L. Kirby, and W. M. Raike (1967). Solution theorems in probabilistic programming: A linear programming approach. *J. Math. Anal. Appl.* **20**, 565–582.

Charnes, A., M. J. L. Kirby, and W. M. Raike (1969). Zero-zero chance constrained games. *In Proc. 4th Intern. Conf. O.R.* (D. B. Hertz and J. Meleze, eds.), *Boston, 1966.*

Charnes, A., M. J. L. Kirby, and W. M. Raike (to appear). An acceptance region theory for chance constrained programming.

Charnes, A. and A. C. Stedry (1964). Investigations in the theory of multiple-budgeted goals. *In* "Management Controls: New Directions in Basic Research" (C. P. Bonini *et al.*, eds.). McGraw-Hill, New York. p. 186–204.

Charnes, A. and A. C. Stedry (1966). A chance-constrained model for real-time control in research and development management. *Man. Sci.* **B12**, 353–362.

Charnes, A. and S. Thore (1966). Planning for liquidity in financial institutions: The chance-constrained method. *J. Finance*, 649–674.

Dantzig, G. B. (1955). Linear programming under uncertainty. *Man. Sci.* **1**, 197–206.

Dantzig, G. B. and A. R. Ferguson (1956). The allocation of aircraft to routes. *Man. Sci.* **3**, 45–73.

Dantzig, G. B. and A. Madansky (1961). On the solution of two-stage linear programs under uncertainty. *Proc. 4th Berkeley Symp. Prob. Statistics*, 165–176.

Dantzig, G. B. and P. Wolfe (1960). Decomposition principle for linear programs. *Operations Res.* **8**, 101–111.

Dempster, M. A. H. (1968). On stochastic programming I. Static linear programming under risk. *J. Math. Anal. Appl.* **21**, 304–343.

Elmaghraby, S. E. (1959). An approach to linear programming under uncertainty. *Operations Res.* **7**, 208–216.

Elmaghraby, S. E. (1960). Allocation under uncertainty when the demand has continuous d.f. *Man. Sci.* **6**, 270–294.

Evers, W. H. (1967). A new model for stochastic linear programming. *Man. Sci.* **13**, 680–693.

Freund, R. J. (1956). The introduction of risk into a programming model. *Econometrica* **24**, 253–263.

Gass, S. I. and T. Saaty (1955). The computational algorithm for the parametric objective function. *Naval Res. Log. Quart.* **2**, 39–45.

Gnedenko, B. V. and A. N. Kolmogorov (1954). "Limit Distributions for Sums of Independent Random Variables" (Transl.: K. L. Chung.). Addison-Wesley, Reading, Massachusetts.

Gonçalves, A. S. (1969). Primal-dual and parametric methods in mathematical programming. Ph.D. thesis, University of Birmingham.

Greenberg, H. J. (1968). On mixed-strategy solutions to chance constrained mathematical programs. Computer Center, Southern Methodist University, Dallas, Texas.

Iosifescu, M. and R. Theodorescu (1963). Sur la programmation linéaire. *C. R. Acad. Sci. Paris* **256**, 4831–4833.

Jensen, J. L. W. V. (1906). Sur les fonctions convexes et les inégalités entre les valeurs moyennes. *Acta Math.* **30**, 175–193.

Kall, P. (1966). Qualitative Aussagen zu einigen Problemen der stochastischen Programmierung. *Z. Wahrsch. verw. Geb.* **6**, 246–272.

Kataoka, S. (1963). A stochastic programming model. *Econometrica* **31**, 181–196.

Lévy, P. (1939). Sur certains processus stochastiques homogènes. *Compositio. Math.* **7**, 283–339.

Lukacs, E. (1960). "Characteristic Functions." Charles Griffin and Co., London.

Madansky, A. (1959). Some results and problems in stochastic linear programming. *Rand Report*, P 1596.

Madansky, A. (1960). Inequalities for stochastic linear programming problems. *Man. Sci.* **6**, 197–204.

Madansky, A. (1962). Methods of solution of linear programs under uncertainty. *Operations Res.* **10**, 463–471.

Madansky, A. (1963). Dual variables in two-stage linear programming under uncertainty. *J. Math. Anal. Appl.* **6**, 98–108.

Mangasarian, O. L. (1964). Nonlinear programming problems with stochastic objective function. *Man. Sci.* **10**, 353–359.

Mangasarian, O. L. and J. B. Rosen (1964). Inequalities for stochastic nonlinear programming problems. *Operations Res.* **12**, 143–154.

Midler, J. L. (1970). Investment in network expansion under uncertainty. *Transportation Res.* **4**, 267–280.

Midler, J. L. and R. D. Wollmer (1969). Stochastic programming models for scheduling airlift operations. *Nav. Res. Logist. Quart.* **16**, 315–330.

Miller, B. L. and H. M. Wagner (1965). Chance constrained programming with joint constraints. *Operations Res.* **13**, 930–945.

Naslund, B. and A. Whinston (1962). A model of multi-period investment under uncertainty. *Man. Sci.* **8**, 184–200.

Prékopa, A. (1966). On the probability distribution of the optimum of a random linear program. *SIAM J. Control* **1**, 211–222.

Prékopa, A. (1970). On probabilistic constrained programming. *In Proc. Princeton Symp. Math. Progr.* (H. W. Kuhn, ed.). Princeton Univ. Press, Princeton, New Jersey.

Radner, R. (1955). The linear team: An example of linear programming under uncertainty. *In Proc. 2nd Symp. Lin. Progr.* (H. A. Antosiewicz, ed.), *Washington*, 381–396.

Radner, R. (1959). The application of linear programming to team decision problems. *Man. Sci.* **5**, 143–150.

Reiter, S. (1957). Surrogates for uncertain decision problems: Minimal information for decision making. *Econometrica* **25**, 339–345.

Sengupta, J. K. (1966). The stability of truncated solutions of stochastic linear programming. *Econometrica* **34**, 77–104.

Sengupta, J. K. (1969). Safety first rules under chance-constrained linear programming. *Operations Res.* **17**, 112–132.

Sengupta, J. K. and J. H. Portillo-Campbell (1970). A fractile approach to linear programming under risk. *Man. Sci.* **16**, 298–308.

Sengupta, J. K., G. Tintner, and C. Millham (1963). On some theorems in stochastic linear programming with applications. *Man. Sci.* **10**, 143–159.

Sengupta, J. K., G. Tintner, and B. Morrison (1963). Stochastic linear programming with applications to economic models. *Economica* **30**, 262–275.

Simon, H. A. (1956). Dynamic programming under uncertainty with a quadratic criterion function. *Econometrica* **24**, 74–81.

Simon, H. A. (1957). "Models of Man, Part IV: Rationality and Administrative Decision Making." Wiley, New York.

Simons, E. (1962). A note on parametric linear programming. *Man. Sci.* **8**, 355–358.

Symonds, G. H. (1967). Deterministic solutions for a class of chance-constrained programming problems. *Operations Res.* **15**, 495–512.

Szwarc, W. (1964). The transportation problem with stochastic demand. *Man. Sci.* **11**, 33–50.

Theil, H. (1957). A note on certainty equivalence in dynamic planning. *Econometrica* **25**, 346–349.

Thomas, D. R. and H. T. David (1967). Game value distributions. *Ann. Math. Statist.* **38**, 243–250.

Tintner, G. (1941). The pure theory of production under technological risk and uncertainty. *Econometrica* **9**, 298–304.

Tintner, G. (1955). Stochastic linear programming with applications to agricultural economics. *In Proc. 2nd Symp. Lin. Progr.* (H. A. Antosiewicz, ed.). *Washington*, 197–228.

Tintner, G. (1960). A note on stochastic linear programming. *Econometrica* **28**, 490–495.

Tintner, G., C. Millham, and J. K. Sengupta (1963). A weak duality theorem for stochastic linear programming. *Unternehmensforschung* **7**, 1–8.

Vajda, S. (1958). Inequalities in stochastic linear programming. *Bull. Inter. Statist. Inst.* **36**, 357–363.

Vajda, S. (1961). "Mathematical Programming." Addison-Wesley, Reading, Massachusetts.

Vajda, S. (1967). Nonlinear programming and duality. *In* "Nonlinear Programming" (J. Abadie, ed.). North-Holland Publ., Amsterdam.

Vajda, S. (1970). Stochastic Programming. *In* "Integer and Nonlinear Programming" (J. Abadie, ed.). North-Holland Publ., Amsterdam.

Van de Panne, C. and W. Popp (1963). Minimum-cost cattle feed under probabilistic protein constraints. *Man. Sci.* **9**, 405–430.

Van Moeseke, P. (1965). Stochastic linear programming. *Yale Econ. Essays* **5**, 196–253.

Van Slyke, R. and R. Wets (1966). Programming under uncertainty and stochastic optimal control. *SIAM J. Control* **4**, 179–193.

Wagner, H. M. (1955). On the distribution of solutions in linear programming problems. *J. Amer. Statist. Assoc.* **53**, 161–163.

Walkup, D. W. and R. Wets (1967). Stochastic programs with recourse. *SIAM Appl. Math.* **15**, 1299–1314.

Walkup, D. W. and R. Wets (1969). Stochastic programs with recourse. On the continuity of the objective. *S.I.A.M. Appl. Math.* **17**, 98–103.

Walkup, D. W. and R. Wets (1970). Stochastic programs with recourse. Special forms. *Proc. Princeton Symp. Math. Progr.* (H. W. Kuhn, ed.). Princeton Univ. Press, Princeton, New Jersey.

Wessels, J. (1967). Stochastic programming. *Statistica Neerlandica* **21**, 39–53.

Wets, R. (1966a). Programming under uncertainty: the complete problem. *Z. Wahrsch. verw. Geb.* **4**, 316–339.

Wets, R. (1966b). Programming under uncertainty: The equivalent convex program. *SIAM Appl. Math.* **14**, 89–105.

Wets, R. (1966c). Programming under uncertainty: The solution set. *SIAM. Appl. Math.* **14**, 1143–1151.

Williams, A. C. (1963). A stochastic transportation problem. *Operations Res.* **11**, 759–770.

Williams, A. C. (1965). On stochastic programming. *SIAM Appl. Math.* **13**, 927–940.

Williams, A. C. (1966). Approximation formulas for stochastic linear programming. *SIAM Appl. Math.* **14**, 668–677.

Žáčková, J. (1966). On minimax solutions of stochastic linear programming problems. *Časopis Pěst. Mat.* **91**, 423–430.

Index

Numbers in italics refer to the pages on which the complete references are listed.